The fast guide to statistical testing with JASP

Dedicated to

Elena Davis

The fast guide to statistical testing with JASP

Classical statistics for social sciences – plus Bayesian tests

Cole Davis

Statistics without Mathematics series

General Editor
Cole Davis

Vor Press

© Cole Davis 2023

First published in Great Britain by

Vor Press

21 Chalk Hill Road,

Norwich NR1 1SL

www.vorpress.com

This book has been deposited with the British Library.

paperback
ISBN 978-1-915500-25-0

hardcover
ISBN 978-1-915500-24-3

ebook
ISBN 978-1-915500-26-7

The right of Cole Davis to be identified as the author of this book has been asserted in accordance with the Copyright, Designs and Patents Act 1988.

All rights reserved. No part of this publication may be reproduced, stored in a retrieval system, or transmitted, in any form or by any means, electronic, mechanical, photocopying, recording or otherwise, except as permitted by the Copyright, Designs and Patents Act 1988, without the prior permission of Vor Press.

While great care has been taken in the production of this book, no claims are made about its accuracy or completeness, and any implicit warranties of merchantability or fitness for a particular purpose are particularly disclaimed. The publisher is not providing professional services, and neither the author nor publisher is liable for any damages in this respect. Should professional advice or expert services be needed, a competent professional should be sought.

Contents

Chapter 1 – Introduction — 1

Chapter 2 – Research design — 11
   Experiments, control groups, variables and other terms 11
   What we are trying to achieve 12
   Data types 14

Chapter 3 – Descriptive statistics — 15
   Central tendency 15
   Dispersion 17
   Assumptions for parametric tests 18
   Testing for distribution 20

Chapter 4 – Null hypothesis significance testing — 22
   One-tailed and two-tailed hypotheses 26
   One Sample T-Test – does a sample belong to a population? . 28
   The two-tailed hypothesis revisited 30

Chapter 5 – Bayesian statistics — 31
   Classical statistics – a brief preparatory overview 31
   Bayesian statistics as the antithesis of classical statistics 33
   Bayesian statistics introduced, via conditional probability . . . 34

Contents

A brief history 34
How are Bayesian statistics used to test hypotheses? 36
And now, even better news! 39
Reporting Bayesian results 40
Three important technical points 43
Which tests to use, classical or Bayesian? 43
A practical example – One Sample T-Test 46

Chapter 6 – Tests of differences   48
Design considerations for the analysis of differences 48
    Some research design terminology applied 50
Tests for same subjects 52
    Paired Samples *t* test: a parametric test for two conditions, same subjects 52
    The Wilcoxon test: a non-parametric test for two conditions, same subjects 56
    Repeated Measures one-way ANOVA: a parametric test for more than two conditions, same subjects (also known as the Within-subjects one-way ANOVA) . 60
    Friedman: a non-parametric test for more than two conditions, same subjects 67
Tests for different subjects 70
    Independent Samples T-Test: a parametric test for two conditions, different subjects 70
    The Mann-Whitney test: a non-parametric test for two conditions, different subjects 75
    Between-subjects one-way ANOVA: a parametric test for more than two conditions, different subjects 77
    Kruskal-Wallis: a non-parametric test for more than two conditions, different subjects 86

Chapter 7 – Tests of relationships   89
Correlations 89

Correlations and effect sizes	94
The Pearson test: a parametric correlational test	96
The Spearman and Kendall's *tau-b* tests: non-parametric correlational tests	101
A cautionary note	106
Multiple correlations – parametric - using Pearson's test	108
Multiple correlations – non-parametric - using Spearman/Kendall's *tau-b*	113
Regression	115
Simple linear regression (two conditions) – parametric	115
Multiple regression – multiple predictors against one dependent variable (parametric)	123

Chapter 8 – Categorical analyses 135

Introduction	135
Speaking categorically	135
The quantification of qualitative data	136
What's happening, statistically speaking?	137
The binomial test: a frequency test for dichotomies (either/or)	138
The multinomial test: a frequency test for more than two categories	143
The chi squared Test of Association: a frequency test for two variables	149
Log-linear regression: modeling three or more categorical variables	161

Chapter 9 – Exercises 165

Questions	165
Answers	167

Chapter 10 – Reporting research 169

Data – absolute or averages?	169
Different audiences	171

Contents

    Graphics 173
    To a live audience!. 174

Chapter 11 – Factorial ANOVA and multiple comparisons   176

    Typical case studies 177
        Factorial analysis of variance – within subjects 177
        Factorial analysis of variance – between subjects 177
        Factorial analysis of variance – mixed design 177
    Effect sizes for factorial ANOVA 177
    Repeated Measures Two-Way ANOVA 178
    Repeated Measures Three-Way ANOVA 185
    Between-Subjects ANOVA 190
    Mixed ANOVA 195
    Multiple comparisons 201

Chapter 12 – A taste of further statistical methods   205

    Data reduction methods 205
        Principal components analysis and factor analysis 205
        Cluster Analysis 206
    Logistic regression 207
    Survival analysis 208
    Reliability 209
        Internal reliability 209
        Test-retest reliability 210
        Inter-rater reliability 210
    Meta-analysis 210
    ANCOVA – some words of warning 210
    Sequential regression – more words of warning 212

References   **214**

Index   **220**

Chapter 1 – Introduction

Is this book for me?

With worked examples from the social sciences, this book is primarily aimed at readers who are new to statistics and are not attracted to mathematics. However, more advanced readers may also benefit from its introduction to Bayesian statistics. Each 'classical' test is accompanied by the same worked example with a Bayesian equivalent.

Short and without any mathematical formulae, the book starts with the assumption that you know nothing about statistics and builds you up slowly rather than hurling a comprehensive toolkit at you. But toolkit it is. While this is very much a book for beginners, you should soon be able to analyze a wide range of common problems.

I want to study social science and they make me do statistics

Apart from being made to do it, I can think of some good reasons for studying statistics. Academically it is a good idea, as you can not only find out if your ideas work in the real world but also demonstrate the extent of their success. It is also likely to stand you in good stead in employment.

Whether you are working in criminal justice, applied psychology, social projects, education, marketing, personnel work, government social policy units or politics, some of the usual destinations for social scientists, if you are known to be a confident user of statistics, you become more of an attractive proposition. You will be able to evaluate projects so that managers can get a better idea of what works and how; interpret surveys so that organizations gain insights into what their clients are thinking; and integrate observations, interviews and focus groups into more obviously quantifiable results.

You might even find, perhaps later in life, that you enjoy research and data analysis and end up as an information specialist. Social researchers are in demand.

Is it really relevant?

The early greats of sociology used statistics. Emile Durkheim wanted to prove that Catholic countries had lower suicide rates than Protestant countries. Friedrich Engels studied mortality rates as part of his highly influential *The Condition of the Working Class in England* (1845). Karl Marx supported his theories by studying the Blue Books, British government records of factory and census statistics. Ferdinand Tönnies was passionate about statistics (as you will be!), studying age- and gender-specific suicide rates, with other factors including distinctions between rural and urban areas. In the mid-20th century, the American Paul Lazarsfeld professionalized social research, and data analysis has continued to be used ever since (c.f. Bickel, 2013 and Savage, 2015).

Psychology is absolutely awash with statistical testing. In fact, so important is measurement in psychology, that many of the ground-breaking tests were created by psychologists. Charles Spearman, whose correlation method you will meet in due course (if you haven't already), resigned from the army in 1897 to take a PhD in psychology. Before he completed it, he had already published a seminal paper on the factor analysis of intelligence and became a pioneer in using mathematical methods within the subject. His predecessor in developing correlations was that pioneer of the study of intelligence, Francis Galton. After the creation of experimental design by R.A. Fisher (a non-psychologist) in the 1930s, tests in psychology proliferated.

Criminology has also gained much from the study of statistics. The positivist thinkers of the 19th century sought for causes of crimes rather than seeing them as gratuitous acts of will. Andre-Michel Guerry and Lambert-Adolphe-Jacques Quetelet used French statistics from the 1820s to demonstrate relevant phenomena still recognizable today: young males in poor neighbourhoods are more likely to commit crimes, but relative deprivation is more causative than actual poverty. They in turn influenced Henry Mayhew and Joseph Fletcher in Britain, whose interest in the problems of urbanization in the 1830s and 1840s anticipated the American experience in the early 20th century. Criminal statistics became increasingly comprehensive in the England and Wales of the nineteenth century and methodology became more systematic. Enrico Ferri, for example, writing in the late 19th century, emphasized the examination of facts, and finding a central idea (Walsh and Ellis, 2007).

Probably the most famous use of statistics was the court evidence of Henri Poincaré in the Dreyfus case, which was important for its exposure of anti-Semitism in late nineteenth French society. From the 1960s onwards, particularly with the advent of widespread computing, criminal analysis became widespread within criminal intelligence agencies. One suggested 21st century usage is the judging of immigrants' ages to check that they are children, using dental studies of molar development (Lucy, 2005); recent doubts have understandably been raised about its accuracy if individuals are subjected to just the one technique on its own, leaving aside the ethics of dentists being called upon to conduct such checks (Bulman, 2016).

All the mathematics you will need

This book is for people who need to use statistical tests but who do not feel comfortable with mathematical explanations. To those who say that only equations express statistical concepts well, I provide the answer given by the publisher of another of my books: "This is for the rest of us" (thanks, Ted). There is no assumption of prior statistical knowledge and no use of formulae. You need no mathematical knowledge other than basic arithmetic, as applied to the decimal system. If you understand the following, you are ready to go:

Differences in sizes within the decimal system; these examples gradually decrease in size, towards zero:
1 .75 .5 .25 .125 .1 .07 .05 .03 .01 .005 .001

Beyond zero, we see negative numbers, with increasing negativity away from zero:
- 0.001 - 0.002 - .005 - .01 -.02 -.05 -.07 -.1 -.125 -.5 -.75 -1 -2

The sign < means 'less than', for example: .005 < .01. The sign > means 'greater than', for example .02 > .01. We use <= to mean 'less than or equal to'; we use >= to mean 'greater than or equal to'.

To multiply a number by itself is to 'square' a number. For example, 0.02 × 0.02, or '0.02 squared', equals 0.0004, a much smaller number. Note that if you multiply two negatives, for example, -.02 × -.02, this should also yield a positive number (here, .0004). Try out your calculator to make sure that it can calculate the double negative correctly. If it doesn't, remove both negatives from your calculations.

In some instances, you will see * for multiplication, e.g. 4 * .05 = 0.2.

That is it. You have enough mathematical knowledge to use this book. You will come into contact with some statistical terms such as mean, median, mode and variance, but even those won't require any formulae or mathematical jiggery-pokery.

What is the teaching strategy of this book?

As well as refusing to use formulae, I have tried to make this book as short as is reasonably possible. I know that there is a feeling of security in having a 1200 page tome that appears to have everything. It is my belief, however, that instructions which convey the most important information are more likely to be understood and remembered. Too much information and the reader is not sure what is of real value. In short, this is not a course on statistics; it is a practical book on statistical testing.

The content is also geared to learning needs. Recognizing that people can only take in so much at one time, the earlier worked examples in any given chapter are simpler, with more information accreting later on. The decision to include reporting in Chapter 10 reflects the likelihood of having to make presentations while studying. More technical matters belong to the last two chapters.

Worked examples are provided for each test. The usual habit of posing questions at the end of each chapter has been eschewed. This is primarily because the real world does not hold your hand and say "Hm, this looks like a job for a Kruskal-Wallis test", but also because there seems little reason to keep the reader under test conditions at all times. On the other hand, to ensure understanding at the level of being able to choose the right tests for basic problems, exercises in the form of simple case studies are available in Chapter 9.

Some of the controversies in the world of statistical testing are discussed in passing. Apart from any entertainment value, this is in my opinion quite necessary. Authors who skirt around awkward choices are not around when the test user is thinking 'well is it me, or does this seem rather strange?' In many cases, the fool is not the test user, who is merely stumbling upon the nub of a debate that has been raging for years outside the cosy world of textbooks.

What's in the book?

The first four chapters provide the basic underpinnings for carrying out data analysis using classical statistics. A fifth chapter introduces

Bayesian statistics; this is very much an optional chapter, but may enrich your understanding of using statistics to examine hypotheses. *

The next three chapters are generally devoted to the statistics that have always been taught at undergraduate level. For many readers, this will be sufficient for quite a long time. Chapter 6 deals with analyses of differences between variables; are the differences in a study likely to be generalizable to the wider world or are they likely to be a fluke? Chapter 7 deals with relationships between variables. The latter part of the chapter includes multiple regression, a technique which used to be neglected in introductory textbooks but is now increasingly used in social research (for example, Bickel, 2013). If you've always wondered about whether or not some of those interviews and focus groups could be quantified in some way, then Chapter 8 is for you. It counts observations, analyzing the frequencies for each category. Most introductory books only cover 'Chi Squared'; this book goes further, as does JASP. Exercises then follow to test your general understanding of which tests are usable in what contexts. There is also a chapter on reporting. This covers both written work and verbal presentation, although it does not teach how to write academic reports: apart from the multitude of works on the subject, this decision reflects the fact that most universities, let alone countries, have rather different expectations as to what exactly should be included in a formal report.

Chapter 11 is an extension of what you will already have learned about ANOVA. Factorial ANOVA involves additional factors, but this does not just add more analyses of differences: relationships between effects also hove into view. The interpretation of visual charts will be one of the particularly important aspects of this chapter. The chapter ends with a section on multiple comparisons. Some university tutors merely say 'use Bonferroni' or 'use the Holm test', leaving the more curious students in something of a quandary. After this chapter, you will not only know about the relevance of particular tests, but will have some understanding of the ethics of their use.

---

*A note on Bayes: Please don't complain if your tutors do not use Bayesian tests when teaching you. This type of statistical analysis has only become computable in the 21st century and is not yet in the mainstream – you're ahead of the game!

Chapter 12 takes you into more advanced territory, but only in summary. Most of the tests related to this chapter are available in JASP, either in its main menu or its easily loaded modules. Some are tests you might use later on in research projects, particularly the data reduction methods and also logistic regression. Survival analysis is more likely to be needed in applied research; the section on this subject gives some idea of the breadth of potential topics, both historical and current. Another less well-covered but very important topic is that of reliability. Also of potential use in later research is meta-analysis. The chapter ends with warnings about two sets of tests that seem to find their way into computer packages, but are not necessarily safe to use: ANCOVA and sequential ('stepwise') regression.

How much do I need to read? Subtitle: How much can I skip?

Beginners

You need to read all of the first four chapters. If Bayesian statistics are not required, you can easily skip Chapter 5.

Ideally you should read all of chapters 6, 7, and 8, following all of the worked examples. Again, however, you can miss out the subsections which bear the title 'Bayesian equivalents'.

You should try to complete the exercises in Chapter 9, although you should have read Chapters 6 to 8 in order to do this.

At some point, try to improve your knowledge of ANOVA by tackling Chapter 11.

If you are in a hurry to use a test without having read all of Chapters 6, 7, and 8, do cover the necessary groundwork. Within Chapter 6, if you want to use tests of differences using more than two conditions (analyses of variance, Friedman, Kruskal-Wallis), you really should study the two-condition tests first (T-Tests, Wilcoxon, Mann-Whitney) – it won't take long. Factorial ANOVA, in Chapter 11, should of course be preceded by one-way ('univariate') ANOVA.

Within Chapter 7, correlations should precede simple regression, and both are necessary before undertaking multiple regression. Log-linear analysis should be preceded by the earlier parts of Chapter 8, particularly the chi squared test of association, and also Chapter 7, especially the section on regression.

You may wish to move on to studying tests cited in Chapter 12 (other books from Vor Press cover most of these in some detail). Of these, study of ANCOVA should be preceded by Chapter 6 and Chapter 7 (and still avoided, perhaps). Factor analysis and/or principal components analysis should be preceded by the whole of Chapter 7, on correlations and regression. Logistic regression should be preceded by Chapter 7 in its entirety.

I would suggest also looking at the chapter on reporting (Chapter 10), especially if somebody wants you to get up and address an audience.

Intermediates and returners

You can skip the first four chapters, although I would recommend revising null hypothesis testing, in Chapter 4, especially on the subject of p values and critical values. If it was all a long time ago, Chapter 3 on descriptive statistics might also be helpful. If it is all terribly familiar, you might entertain yourself by reading Chapter 5, on Bayesian statistics.

If there are tests that you have not used before, follow the advice for beginners, 'If you are in a hurry'. Obviously, one should build upon accumulated knowledge.

Why use JASP?

It is free and open source. While data input is similar to SPSS, its expensive rival, procedures have been streamlined – note in particular the comparative simplicity of setting up a repeated measures ANOVA –

and options can be altered with ease. JASP also contains effect sizes and confidence intervals, of increasing importance in modern statistics.

JASP has a core of tests that are usable within a typical undergraduate course, but the package offers additional tests, in easy to load modules, that may offer a broader learning experience. One key focus of JASP is reproducibility; options chosen by the user may be saved within a file. JASP has its own editing style: you can double-click and edit your preferred spreadsheet, with saved data automatically updating the analysis.

A brief note on data entry

JASP supports a range of files: those with separated columns such as .csv and .txt; SPSS, SAS and Stata files; .ods (open document spreadsheets as in LibreOffice and OpenOffice).

Here, I open the file **Basic.csv**, with just 10 cases (files can be downloaded from the publisher's website).

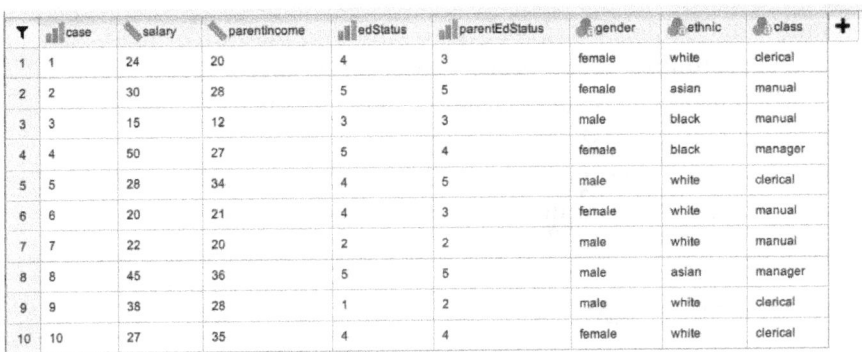

Each observation should have its own row. In a same-subjects design, say looking at the relationship between salary and parental income, the relevant variables are used in a straightforward manner. In a between-subjects design, such as examining differences in educational background among classes, then a grouping variable such as 'class' is used to examine a metric such as 'edStatus'. This will become very familiar to you as you read the book, which has worked examples for each test.

Details about data handling may be found on the JASP website.

Acknowledgements

I am indebted to the following contributors for their advice, encouragement and cooperation at different times in the gestation of the Statistics without Mathematics series: Casper Albers, Michy Alice, Ed Boone, Iain Buchan, Winston Chang, Michael Fay, Henriette Hogh, Jonathon Love, Sharon McGrayne, Abraham Mathew, Boris Mayer, Richard Morey, William Revelle, Grant Schneider, David Speed, Marianne Vitug, E-J Wagenmakers and Douglas Wolfe.

I am also grateful to previous commissioning editors who have provided useful criticism and support for my ideas on teaching statistics to complete strangers. In approximate chronological order, these were Karen Winter (née Bowler) at Policy Press, with John Manger and Ted Hamilton at CSIRO. Also, I appreciated the stimulation from discussions with their counterparts at John Wiley and Routledge.

The additional assistance and further guidance of Ted Hamilton, publisher emeritus, is deeply appreciated.

I apologize if I've missed anybody or failed to take up advice as proffered. Any errors or questionable judgements are my own.

Feedback request

If you wish to submit reviews, make suggestions for improvements or point out errors, please contact me via the publishing website.

Cole Davis 2023

Chapter 2 – Research design

Experiments, control groups, variables and other terms

There are plenty of books on research design, but this book is primarily about statistical testing, particularly using JASP. This being the case, we will only consider the basics, as they pertain to testing. As you go through the examples in the book, these terms will become more familiar and easier to understand.

The terms **independent variable** and **dependent variable** will be referred to regularly. These are respectively the variables being manipulated and the variables affected by such manipulation. You will also come across the terms **predictor** and **criterion**, the former variable influencing the situation and latter variable being the item of measurement. The two pairs of terms are interchangeable in much of the literature. Strictly speaking, **experiments** should refer to independent and dependent variables, as shown in this example:

One group of social work managers, randomly chosen, receives additional training in how to provide feedback to their staff; this is the **experimental group**. Another group of social work managers is not given training; this is the **control group**.

The difference between the two **conditions** (in data analysis, often referred to as **levels**) is judged by successful outcomes over the next six months, or levels of problematic incidents or whatever measure is considered most suitable. The independent variable (sometimes abbreviated to **iv**) is the existence or otherwise of feedback training; the dependent variable (**dv**) is performance.

We could have a **quasi-experiment**, perhaps using past records to see how feedback affects performance, or maybe using naturally occurring groups – for example, one local authority has this training and a similar authority does not, and we contrast the groups. Here we do not really have a control group. Strictly speaking, we should not use the terms 'independent variable' or 'dependent variable'. The use or otherwise of training is the predictor, with the different levels of performance as the criterion (or 'measure'). However, you will see both sets of terms used in the literature, regardless of how the research is conducted.

Generally speaking, however, we use the terms predictors and criteria in regression. For example, we may be interested in the effect of different media portrayals of gender on children's behavior. The watching of internet clips and computer games may be the predictors, whereas levels of gender-specific behavior may be the criterion.

What we are trying to achieve

Some of the time we are trying to look at *differences* between conditions. This will be seen in particular with t tests, ANOVA, and their non-parametric equivalents. There is a further sub-divide, whether or not you use a **between subjects** design, also known as an **unrelated design**, where different people are tested in each condition, or a **same-subjects** design (also known as **within subjects** or **related design**). The advantage of using the same subject under the different conditions is that the **effect** under study is unlikely to be conflated with individual differences. Often, where we cannot use the same person, we may try to **pair** (or **match**) the subjects (participants, if human beings) in the relevant areas; for example, if we wanted to contrast two types of violent media, but wanted each child to only see one of them, we could choose children to be paired according to similar levels of tested intelligence, age and social background, making them similar for the purpose of the experiment. **Paired tests** are used for both paired and same-subjects research. For obvious reasons, the same number of subjects is required in each condition.

Sometimes we are unable to use the same subjects. Perhaps the study would be adversely affected by participants experiencing more than one condition, or different subjects are the whole point of a study (males and females, in a gender study, are usually different). Then **unpaired tests**, known in JASP as **independent samples** tests, try to take into account individual differences. There may be different numbers of subjects in each condition.

At other times, we are interested in the *relationships* between conditions. In particular, we examine **correlations**, linear relationships between conditions, positive or negative. For example, we may study a range of different attributes to see if they are inter-related. Perhaps the higher socio-economic status of a person, the more acquaintances he or she has, perhaps the easier it is to meet other people. It should be noted that correlations do not necessarily demonstrate cause and effect. For example, some theorists may suggest that gregarious personalities are more likely to climb socially and economically, rather than their status affecting their opportunities for social mobility.

A correlation requires the intersection of informational pairs, for example with each individual's scores on one measure matched with their scores on another. For this reason, there must be equal numbers in each condition.

Data types

Another issue is the nature of the data. The above situations generally require measurable data, but the choice of test depends upon the granularity of the data. **Continuous** data is proportional and takes on a 'natural' feel, like a range of body weights, times and ages, 48, 50, 53, 55, 55, 56, 58 and so on. **Ordinal data** can include the results of an uncalibrated Likert scale (1 to 5, 1 to 7, 1 to 9 and so on) or rather coarsely grained, 'lumpy' data, such as 2, 6, 55, 55, 109. The implications will be discussed in Chapter 3.

Other data are counts of observations or incidents: **frequency**. For example, you may want to look at incidents of discrimination within a particular locality over a period. You could find 30 cases of discrimination in the workplace, 10 such cases in recruiting, and 5 in the provision of services, and so on. As well as such differences, you may have matrices of cases allowing a study of relationships. If we have also classified the cases into the different social classes of the victims, we could examine the relationship between the class of the victim and the type of discrimination committed against them.

Frequency counts - 10 blue collar workers, 22 clerical workers, 7 managers - are **categorical** data (also known as **nominal** or even **qualitative** data). There are no quantitative comparisons such as averages; all differences are qualitative. Like elephants and lamp posts, you don't usually consider each individual on the same scale; elephant number 3 and lamp post number 3 cannot be considered for their relative luminosity or suitability for climbing (I think).

Chapter 3 – Descriptive statistics

Central tendency

Given a set of figures, whether or not we compare it with another set of figures, we need some way of representing it. We may want the maximum and minimum values, but when it comes to statistical testing, we are usually more interested in central tendency, basically a representative value which is deemed to be typical. The measure of central tendency is usually the **mean**, the **median** or the **mode**.

If we take this very small data set, 2 3 3 4 8, we can demonstrate the differences between the three measures of central tendency. Here we work them out by hand; we won't need to do it again!

The mean, usually what is meant by 'average', is calculated by dividing the sum of the variable by the total number of cases. The sum here is 20, the number of cases 5, so 20/5 equals a mean of 4.

The median is the mid-point in the range. Calculating the median requires moving to the outer limits, eliminating the values there and continuing until we reach the middle. So in this example, we first eliminate 2 on the left and 8 on the right, then rule out 3 on the left and 4 on the right. Our median is 3. (The even-numbered set 2 3 3 4 4 8 would have 3.5 as its median; remove 2 and 8, then 3 and 4, leaving 3 and 4 in the middle.)

Central tendency

The mode shows the most common response. The mode for 2 3 3 4 8 is of course 3. It is possible to have multiple mode values.

In practical terms, these measures have varying utility. Let us say that we are considering salaries. The strength of using the mean is that it takes into account everybody's income, from the stratospherically well-paid to the lowliest wage-earner; on the other hand, this can be a weakness, say if two or three billionaires distort our figures. The mode may counteract this effect, as it tells us the income of the largest number of people, perhaps of administrative workers; but this is hardly representative of the workforce as a whole. The median gives us a central value, perhaps that of a middle manager; it is useful, but does not take into account the number of people who are wealthy or poor. At this stage, I will merely say that the example shows the need to adapt interpretations to context; no magic button just tells you everything.

As button-pressing is nevertheless quite enjoyable, I suggest opening the **Basic.csv** file and pressing JASP's Descriptives tab. In order to look at some descriptive statistics, transfer the four numerical variables from the left to the Variables box.

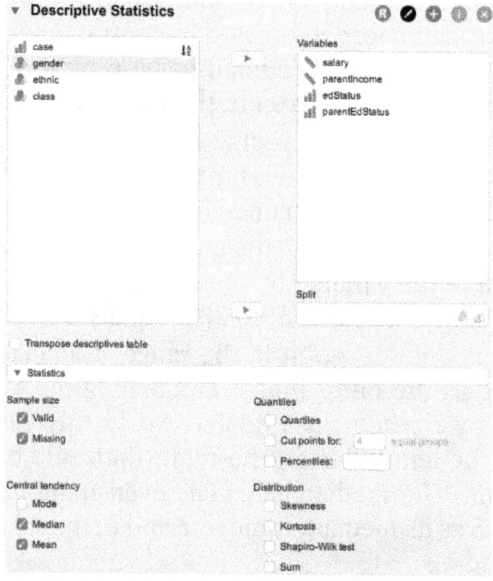

Dispersion

Within the Statistics panel, there is a range of Central tendency options, from which I have added the Median statistic to the default settings.

Descriptive Statistics

	salary	parentIncome	edStatus	parentEdStatus
Valid	10	10	10	10
Missing	0	0	0	0
Median	27.500	27.500	4.000	3.500
Mean	29.900	26.100	3.700	3.600
Std. Deviation	11.190	7.795	1.337	1.174
Minimum	15.000	12.000	1.000	2.000
Maximum	50.000	36.000	5.000	5.000

On the right-hand side is the display screen. As the Likert scales for edStatus and parentEdStatus have 5 points, the median figures reflect this, with a rather blunt 4 and 3.5; although the median is theoretically the best figure when dealing with non-parametric tests (of which, more later), it is often better to use the mean because of a more easily distinguishable spread of values.

In the Statistics section, you will also find such options as the mode and sum, which adds up all of the scores. You can also place a grouping variable into the Split box to see the differences; in the case of 'class', this means seeing the scores for different levels.

Dispersion

Dispersion is how the data is spread. By default, JASP shows us the minimum and maximum values, which are useful for checking for outliers and errors in data entry, and the standard deviation, which shows how spread out the data is; see the distribution chart a little later. The **range** is the difference between the minimum and maximum.

Assumptions for parametric tests

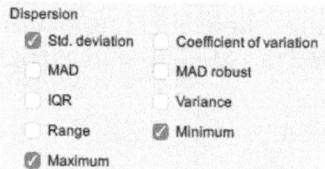

We probably do not need to see the **variance**, but the concept is important. Many of the tests that we will use are related to measures of central tendency. Parametric tests are in particular concerned with variability about the mean.

Assumptions for parametric tests

In general, parametric tests are considered to be more powerful than non-parametric tests and are therefore the weapons of first resort. On the other hand, if the data set does not meet the assumptions for parametric tests, parametric tests may come to conclusions about an imaginary data set; in such a case, the non-parametric test is preferred. A non-parametric test ranks the data, as if they were all ordinal, and makes no assumptions about the distribution.

Unfortunately for test users, there are disagreements among statisticians about the extent to which these assumptions are necessary (this is not the only instance of mathematically inspired fisticuffs). More traditional test users insist on strict adherence to the assumptions for using parametric tests. Others note the robustness of parametric tests and are inclined to be rather less stringent about the assumptions.

We will consider each assumption in turn, looking first at the traditionally taught view and then at some rather more relaxed practices. Before you get too worried about this, you will find that employers and academics will have their own views on the subject, so be prepared to render unto Caesar.. *

---

*My own take on non-parametrics is that when using data considered suitable for parametric tests, they generally show the same results. When the data are unsuitable, non-parametric tests may produce a more conservative result, with good reason.

Assumptions for parametric tests

One assumption is that the data are continuous. There is general agreement that the most lumpy data sets – stuff like 2 8 7 16 316 32 96 – are really unsuited to parametric tests (although they are common in many research studies). Beyond this, agreement tends to go out of the window. The more conservative test user asks, "if you halved this data item, would the new value really be 50% of the old?" Think about, for example, the notion of a service being 'Ok', 'Average', 'Good', 'Very Good' and 'Superb' on a scale from 1 to 5; would halving Very Good mean Average?

When using a Likert scale, the conservative user would subject it to a parametric test only if the scale had already been calibrated in a pilot project.* We do not have room to discuss calibration in this book, but you would do well to look up 'item response theory'. More relaxed test users use parametric tests on Likert scale results in all circumstances.

Another assumption is homogeneity of variance. If you have one set of data that looks like 3 4 7 9 and another that looks like 32 38 52 67, then parametric tests are not suited to working with them both together.

A third assumption is a normal (also known as Gaussian) distribution.

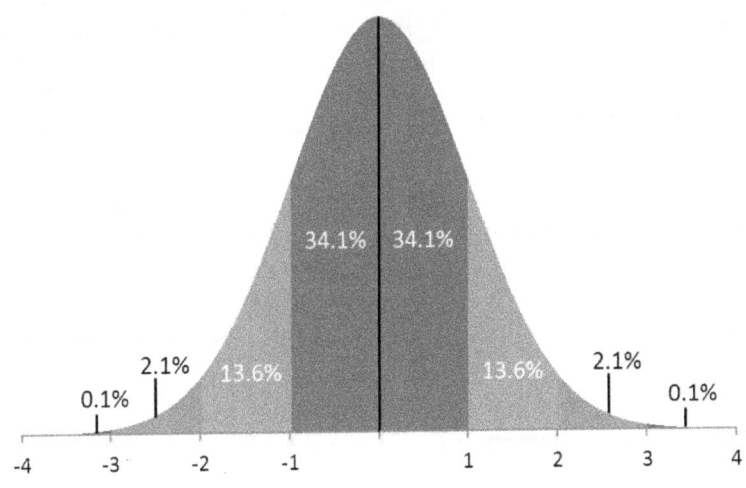

---

*Pilot projects, by the way, are always advisable, as they give you the chance to iron out unexpected problems.

This is an idealized version of the normal distribution. Most of the data is near the central tendency, with less and less of the data lying at the extremes. The measures of dispersion on the axis at the bottom of the chart are standard deviations.

The implication is that your sample is representative of a population, an important concept when we consider null hypothesis testing in Chapter 4. Do note that a population does not have to be a population in the sense of all the people in a geographical area (although it can be). The population may be psychology students in the USA, the unemployed in London or libraries in Australia; the sample is assumed to be representative of that population.

Testing for distribution

In order to decide whether or not we have a distribution which is suitable for parametric test usage, we usually use an heuristic, a practical method which while not guaranteed to work, usually does the trick.

With reasonably large data sets, you can use the histogram and density charts within the Basic plots section to examine the distribution. Small data sets are unlikely to get an approximation of the normal curve shown above. This leads some traditionalists to recommend non-parametric tests for small data sets, although others will live with parametric tests if they meet fairly strict assumptions.

Returning to the Statistics section within JASP's Descriptives, go to the Distribution options and select Skewness and Kurtosis. Skewness is how far the data is spread away from the mean in one direction or the other. Kurtosis is more to do with the shape of the curve, in particular the weight of its 'tails'. For samples of less than 50, consider Skewness and Kurtosis readings of within 1.96 (or -1.96) to be acceptable; although more liberal readings go from +2 to -2. For 50 to 300 cases, the limit should be 3.29 (Kim, 2013). For samples larger than 300, use the Distribution plots option within the Basic plots panel, also accepting a maximum of 2 for Skewness and 7 for Kurtosis (West *et al*, 1995).

Testing for distribution

A more modern approach to testing for normality is to use the Shapiro-Wilk test (Razali and Wah, 2011). Within the Statistics panel, find the Distribution statistics, and select Shapiro-Wilk test. A significant result using this test indicates a pattern that is probably not from a normal distribution. So you want a non-significant result from the Shapiro-Wilk to continue using a parametric test. *

We turn to the notion of significance, or perhaps more strictly, non-significance, in Chapter 4.

---

*The Shapiro-Wilk test within the Descriptives tab is preferred to versions used within specific statistical tests in JASP, which at times can produce rather inconsistent results. If in doubt, return to the Descriptives!

Chapter 4 – Null hypothesis significance testing

Null hypothesis significance testing (NHST) particularly features in classical statistics (also referred to as 'frequentist' statistics by Bayesians, who will turn up soon). As it is the most commonly taught approach, it is particularly important to understand it.

I will keep this fairly short; your part of the deal is to return to this chapter whenever you feel less than clear about the subject. However, you should get more of a feel for it when you have gone through some of the worked examples in the book.

Let us say that the ratings for service A are higher than the ratings for service B, but we are not sure if the difference (the **effect**) is a meaningful one. Maybe industrial performance appears to be related to political turbulence, but we want to know whether or not the relationship (again, the effect) is merely subject to chance.

In NHST, the statistical test examines the **sample** – the cases for which we have evidence – and considers it within the likely **population** from which the sample is drawn. This is why this area of statistics is often referred to as **inferential statistics**: it makes inferences from the data in the sample about the population as a whole. So we may study the behavior of a sample of 30 sociology students, assuming that they are reasonably representative of the population, perhaps sociology students nationwide.

The default position assumed by a statistical test is the **null hypothesis**, sometimes represented as H0. The null hypothesis is that the findings do not differ significantly from chance, noise or experimental error. More prosaically, the null hypothesis says by default that your beloved effect is just garbage. The test's essential role is to tell you whether or not it is reasonable to reject the null hypothesis; that your effect is not random variance. A scientific principle is being maintained, that of falsifiability: according to Popper (1968), in science one can only falsify a theory. We can find out if the null hypothesis can be upheld, yes, but we can't make a direct claim for an effect.

The hypothesis that you are trying to prove in your study is called the **alternative hypothesis**, or H1, or the 'maintained' or 'research' hypothesis. This rejects the null hypothesis. A test only allows you evidence to support the rejection of the null hypothesis, to say with some confidence that the effect is unlikely to be a fluke.

Put another way, if the null hypothesis is rejected, then you can feel that there is some indirect evidence to support the alternative hypothesis. So, hypothesis testing does not directly support the effect under investigation; it merely attempts to disprove the null hypothesis, that your result is a fluke of some type. Saying that a result is 'significant' is something of a lay term in classical statistics; you might use it in applied research, but not under the eyes of your tutor! (Do note that the tests themselves are examining the null hypothesis, that there is no peculiar variance. Unlike you, the computer does not care about your cherished effect!)

The statistic that we most often read in classical statistics to see if the null hypothesis may be rejected is the ***p* value**. This is a decimal number between 0 and 1 which your test will generate. The smaller the number, the more likely it is that we would declare 'significance' (that the null hypothesis can be rejected). So, to give two more or less random examples, $p = .783$ is a large value whereby the results are almost certainly useless for experimental purposes; there is emphatic support for the null hypothesis. Nearer to the other end, $p = .007$ is really quite small (we've been waiting for you, Mr Bond) and, in most cases, we would feel justified in rejecting the null hypothesis.

Well that's all right then. We know that big p values mean that our effects are, well, ineffectual (the null hypothesis again) and that small numbers mean that we are famous.

But this raises a question or two. How big a number does a p value have to be to mean that our effect is a fluke? And are there different levels of what we might call 'small'? And why was I such a nuisance as to say that we would feel justified in rejecting the null hypothesis with p = .007 only "in most cases"?

Enter the **critical value**. It is possible that an experimenter may choose to pre-set a value at which the null hypothesis would be rejected. The typical critical value in social science experiments is $p < .05$ (p is smaller than .05) and it is quite likely that you will tend to use this in your course. So a p value of .046 would allow us to reject the null hypothesis (victory is ours), and a p value of .06 would not. This was suggested by the statistician RA Fisher as a useful rule of thumb, all other things being equal:

> ... it is convenient to draw the line at about the level at which we can say: "Either there is something in the treatment, or a coincidence has occurred such as does not occur more than once in twenty trials". Fisher (1926)

'Once in twenty' is of course the same as 5 in 100 (.05), but please do not treat these proportions as real by putting them into calculations or claiming that you have 95% likelihood or anything like that – they are only probabilities.

There are times when an experimenter would like to set a lower critical value. Commonly seen critical values are $p < .01$ and $p < .001$, although others are also possible. The lower the critical value, the smaller the p value required to support rejection of the null hypothesis. When it comes to aviation safety, for example, I would hope that a critical value as high as $p < .05$ would not be set for testing a mission-critical piece of equipment.

The critical value, particularly at the level $p < .05$, has come under considerable criticism, particularly in the area of psychology. It should be remembered that failure to find a small enough p value does not mean that an effect definitely does not exist. Also, the opposite is possible, that what appears to be an effect is in fact a fluke or the effects of other variables, statistical noise if you like. There is such a thing as the **Type 1 error**, believing that an effect exists when it doesn't; this is why you should be careful not to run too many tests within a study, as some results are likely to be flukes. **Type 2 errors** are the opposite, rejecting an effect which does in fact exist; this can be the effect of being too dogmatic about a cut-off. Either type of error is possible as a consequence of using the wrong test.

However, the continuing usage of the $p < .05$ rule of thumb over so many years does suggest a history of effectiveness in picking up effects (Bross, 1971). Assuming that we accept $p < .05$ as a useful general guide, we are still left with questions such as 'do we reject a p value of .052' and 'is a value such as .045 always a meaningful effect?' Also, is a result significant but not of much use to the world? If you work with 'big data', you will find lots of very small p values and will wonder what to do with them all.

This brings us to another statistic, the **effect size**. This tells us how much of the variance in the sample is likely to be because of the effect. If, for example, you get an effect size such as .671, we can say that the effect is likely to be responsible for 67% of the variance in the sample. At other times, you may feel able to reject the null hypothesis, but find an effect size so small as not to be particularly useful in the real world.

Now, there is nothing wrong with reporting the test statistic, and the p value, and if possible the effect size, and letting the reader decide if the results are acceptable. In fact, I would advocate it, except for the fact that tradition and custom, particularly among publishers of research, has made things much messier. How small a p value is small enough to reject the null hypothesis and report our experiment to the world? Does the publisher insist on neat tables, with p values accompanied by asterisks and critical values (* = $p < .05$, ** = $p < .01$, and so on)?

Regardless of your feelings about citing a critical value, you could publish the actual p value alongside it and indeed the effect size. Do not worry about these terms, as you will see various examples of NHST in action, which should accustom you to the concepts and practice.*

One-tailed and two-tailed hypotheses

I'm sorry, but there is another related concept that has an effect on how we consider our data. If you have good reason to know the direction of an effect before running the test – we are for example quite sure that the mean of A will be smaller than B (but not necessarily how much smaller) – then you may choose a one-tailed hypothesis. A good reason is a clearly explainable rationale or theoretical underpinning for the prediction. Otherwise, you should opt for a two-tailed hypothesis.

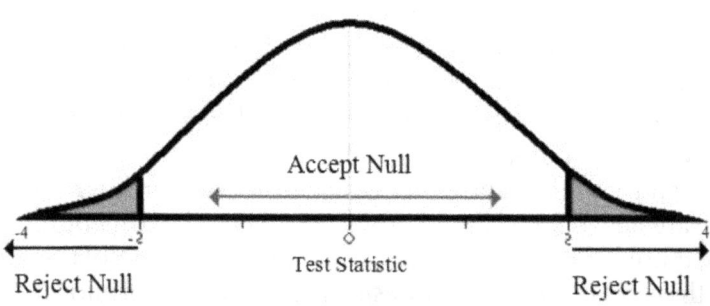

As you will see from the chart, the result must be more rigorous for a two-tailed hypothesis. By allowing for potential variance on either side of the mean, we are having our cake and eating it. This also means a more rigorous approach to the p value, which will be bigger than if a one-tailed hypothesis had been chosen.

*Also sometimes demanded are confidence intervals, to be discussed later in the book.

One-tailed and two-tailed hypotheses

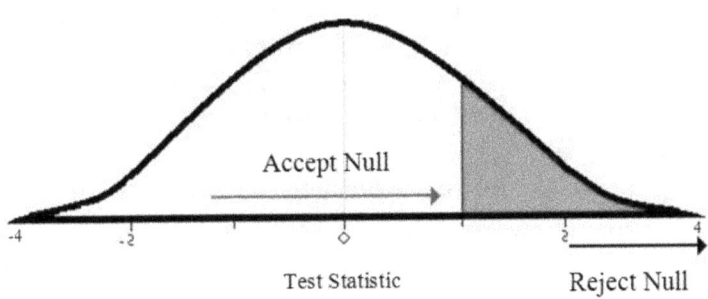

Less variance is under consideration when we are only considering one tail of the distribution curve. This being the case, a more lenient attitude is taken to the results. In practice, this means a reduction in the p value, so it is easier to reject the null hypothesis.

If this is used inappropriately, it is possible to end up with a Type 1 error, wrongly claiming significance. If in doubt, use a two-tailed hypothesis. One thing that you definitely should not do is to opt for a one-tailed hypothesis just to 'find significance'. This is frowned upon.

The following test is one that you probably won't use much. Its main purpose here is to allow you to accustom yourself to the software and to apply some of the concepts discussed.

One Sample T-Test – does a sample belong to a population?

This test is designed to test whether or not a sample belongs to a population when we know the mean of that population. In this case, I believe that salaries should have increased since the previous records and want to see if this is a reasonable claim. I know that the previously recorded mean was 22,000 and I want to find out if current incomes (in thousands) really are bigger.

Having opened the **Basic.csv** file with JASP, we press the T-Tests tab and select Classical / One Sample T-Test from the drop-down menu.

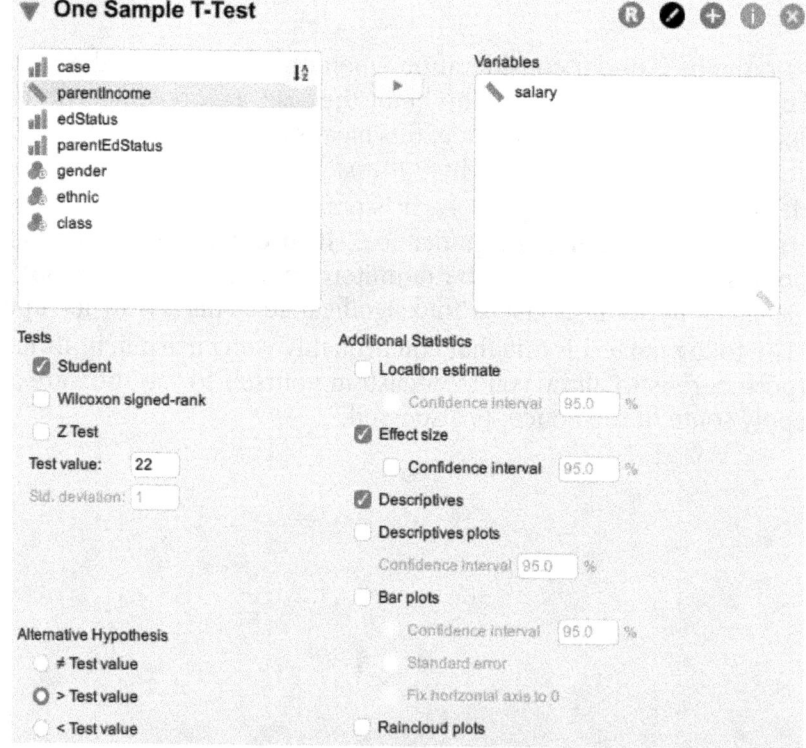

One Sample T-Test – does a sample belong to a population?

I have transferred the Salary variable to the right-hand box, which is where the action takes place. Most importantly, I have entered 22 as my Test value (click anywhere on the screen in order to update the results). I have also gone for a one-tailed hypothesis, stating that the data is greater than the test value. I have also selected Effect size and Descriptives.

One Sample T-Test

	t	df	p	Cohen's d	SE Cohen's d
salary	2.233	9	0.026	0.706	0.353

Note. For the Student t-test, effect size is given by Cohen's d.
Note. For the Student t-test, the alternative hypothesis specifies that the mean is greater than 22.
Note. Student's t-test.

Descriptives

	N	Mean	SD	SE	Coefficient of variation
salary	10	29.900	11.190	3.539	0.374

This is the full display of results on the right-hand side of the program. The T-Test result shows us a p value of .026. We have reason to reject the null hypothesis. We have quite a large effect size (Cohen's d). Looking at the bottom, we get the descriptive data we are most likely to report, in particular the Mean for the data and the standard deviation.

We can check normality with the Shapiro-Wilk test (as suggested before, the version in Descriptives is to be preferred). If Shapiro-Wilk's p value is large, it seems reasonable to assume a normal distribution, and therefore Student's test, a parametric test, is fine to use.

If the Shapiro-Wilk test had been significant, with a very small p value, probably < .05, I would have chosen the 'Wilcoxon signed-rank' test instead, as this is the non-parametric equivalent of the Student test.

The two-tailed hypothesis revisited

Let us return briefly to the two-tailed hypothesis. If I had no theoretical reason to believe that the new data set was different in a particular direction from the test value, I would have opted for the Hypothesis at the top, merely that the data is different from the test value. This might have been because I was uncertain, or perhaps that I had expected the salaries to be approximately the same and wanted to test this; a non-significant result would indicate no substantive change.

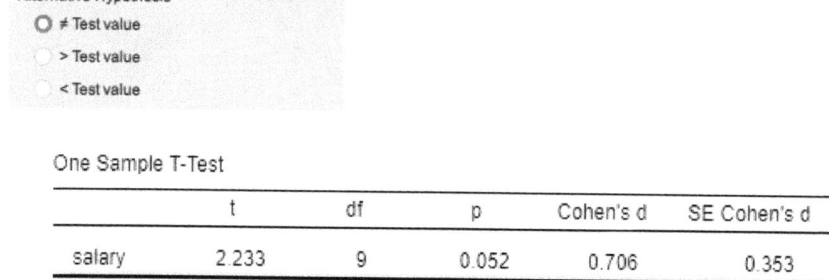

One Sample T-Test	t	df	p	Cohen's d	SE Cohen's d
salary	2.233	9	0.052	0.706	0.353

Adopting the two-tailed hypothesis in our worked example, the p value rises to .052, not quite within Fisher's suggested limits.

Note that in proper research, we should not test on the basis of two incompatible hypotheses; we should decide on our hypothesis before running a test. If, however, you are still in the grip of methodological curiosity, you can see a Bayesian analysis of these situations at the end of the next chapter. (Don't worry if you know nothing about Bayesian statistics. You will by the time you have read the next chapter!)

Chapter 5 – Bayesian statistics

Classical statistics – a brief preparatory overview

Without yet going into details, we will look at some of the differences between the two types of statistics. Up until now, we have been considering classical (sometimes known as 'frequentist') statistics, in which a sample is examined in the expectation that its characteristics will be replicated in the overall population.

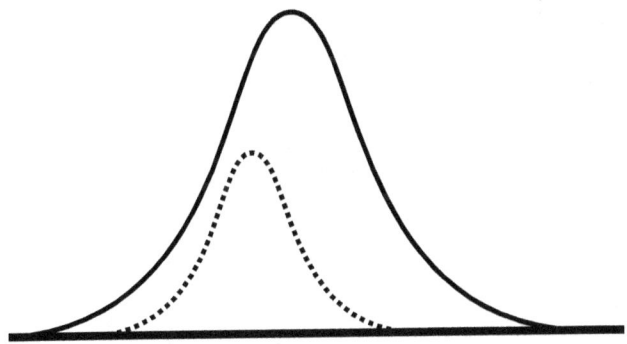

The basic assumption is that if one looks at sample after sample (hence the term 'frequentist'), each would appear quite similar.

Classical statistics – a brief preparatory overview

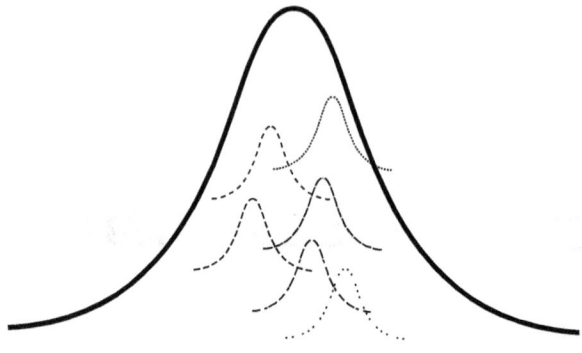

Related to this is the idea of a stopping rule. Not to go into too much fine detail, how would you feel if I said "well the experiment didn't prove my point, so I'll get more observations?" With your knowledge of classical statistics, you would probably think 'what an old fraud!'

Another point about classical statistics is that a large part of it is about rejecting the null hypothesis. While this is very much part of the Popperian view of science, that theories should be disprovable, it is nevertheless a touch convoluted and in statistics does not provide a precise negative: a test does not prove that your effect does not exist; it shows that a random or otherwise neutral situation cannot be disproven.

Yet another point is that 'significance', more respectably the rejection of the null hypothesis, seems pretty much an arbitrary affair: it is, or it isn't. Now when you think about it, this is a peculiar state of affairs. Let's take a *reduction ad absurdum* example. Suppose we have some reason to believe that over time, the absorption of literary fiction creates more considerate, less violent people. While the p value is a perfectly reasonable thing to quote, giving you some idea of the evidence against the null hypothesis, at some point the reader wants to know, "well is it significant or not? you know, at $p < .05$?" (or whatever is the critical value). Are my children going to have more empathy if they just read a bit more Dickens and Annie Proulx? Put at its worst (and if we ignore issues such as effect size and the pragmatism of RA Fisher), there seems to be an absolutism about classical statistics: it's significant or it's not.

Bayesian statistics as the antithesis of classical statistics

I have rather moved into devil's advocate role. Again without going into detail, let us see in what ways the Bayesian approach to statistics can be viewed as a radical alternative to the classical tradition.

Bayesian statistics is not based on the relationship between a sample and a population. The frequentist population is a hypothetical one: you have to guess that your population is uniform enough to allow predictive sampling.

Bayesian statistics is based on probability, dealing with the evidence to hand. The relative unimportance of sampling means that there is no 'stopping rule'. If the population is irrelevant, so is the size of the sample. Let us imagine for example that in classical statistics you need to have 2000 cases for the purposes of a survey; after you have the sample, you study the relevant effects. One of the advantages of the Bayesian method is that you do not need to have the whole sample before you start. If you have, say, 300 cases, and it becomes obvious that the effect under consideration clearly doesn't exist, then you can stop. If you find that there is a tendency towards its existence, then you can gather more evidence.

Evidence is gathered with a much more straightforward logic than in classical statistics. You have evidence in favor of the null hypothesis and you have evidence in favor of the alternative hypothesis, the effect.

As the evidence accumulates, Bayesian statistics does not give us the 'significant or not' dichotomy. It provides a graded view of the evidence: how much is there? You will find in the reporting tables to be discussed shortly that there is something of an equivalent to 'non-significant', but the evidence itself is measured.

Bayesian statistics introduced, via conditional probability

Bayesian statistics considers probability in terms of calculated likelihood, working out how likely it is that something will happen again. Basic examples of what I have in mind are working out how many balls of a different color are likely to be in a bag when one is removed; how likely it is that a person is going to be affected by an illness; the odds of a sports team winning when we know the strengths and weaknesses of the team and its opposition.

Well, that is conditional probability, the probability of an event after another event has taken place. For example, the removal of a red ball from a bag affects the probability of how many balls of different colors are left in the bag. This is also called inverse probability.

By itself, inverse probability has a place in predicting, for example, the likelihood of a particular disease affecting people. It can even work out the probability of one particular set of people getting the disease.

However, Bayesian statistics gives conditional probability a twist in direction. Instead of seeing an actual event and calculating its effect on future events, Bayes is interested in looking at a rather uncertain situation first and then examining the effect on probability when new, known, information is added to the mix. Initially, we take a guess (of which more later), then we update our knowledge with fresh information (the sample) and see how it compares with our prior knowledge.

A brief history

Before we get involved in the details and terminology, it is worth having a quick glimpse at the history of Bayesian statistics. That way, you can see its applications, often vital in modern history, how it works, and that very important point, that it does work.

Thomas Bayes, a Presbyterian minister in eighteenth century Kent, England, took an interest in inverse probability. The preoccupation which led to having a branch of statistics named after him is the idea that probability is to some extent based on a type of belief, rather than hard knowledge in the form of frequencies. We start with some prior knowledge which forms a basic but incomplete theory of events. Then we add evidence.

For some reason, Bayes discontinued his study of this area of mathematics. * It was only after his death in 1761 that his friend Richard Price edited his notes and had them published. However, the first major use of this theory was by the remarkable French scientist Pierre-Simon Laplace, who apparently independently of Bayes, applied inverse probability to interplanetary movement, court testimony and other fields. It may be more accurate to call Bayesian statistics Laplacean, but it was perhaps a mark of his breadth of achievements that the French Isaac Newton abandoned inverse probability in favor of other scientific interests.

Bayesian statistics fell into disuse among academic statisticians but continued to be applied to a range of problems. In the famous Dreyfus case, the principle of updating knowledge with fresh knowledge defeated the collation of coincidences as evidence:

> "Since it is absolutely impossible for us [the experts] to know the *a priori* probability, we cannot say: this coincidence proves that the ratio of the forgery's probability to the inverse probability is a real value. We can only say: following the observation of this coincidence, this ratio becomes X times greater than before the observation." (Darboux *et al*, 1908)

It was also, among other applications, used in American actuarial mathematics, the statistics of smoking and lung cancer, and during the Second World War, Alan Turing's cracking of the Enigma code and the detection of the location of German U-boats. A particularly graphic example of Bayesian methodology was the use of a grid system in searching for a sunken American aeroplane carrying a nuclear bomb; as the search went on, such knowledge as was gathered, by eye-witness

---

*While his thought experiments refer to a table, his use of a billiards table is almost certainly mythical.

testimony and fragments, was built upon by subsequent searches of grid cells. Each patch of the sea was evaluated as somewhat more likely to be positive ('getting warmer' in the hide-and-seek sense) or negative ('getting colder').

Instead of the previous tale of a theory falling into disuse in academia, Bayesian ideas were fiercely opposed through much of the twentieth century, only to emerge as the new rage in the early twenty-first century. For an engaging narrative covering the fearsome rows – often fought by very awkward personalities – as well as the applied successes mentioned above, read McGrayne (2012).

How are Bayesian statistics used to test hypotheses?

To return to Bayesian statistics, we have our prior theory of the world, be it a guess, an expert opinion or no theory whatsoever, and we then combine this theory with the results of the study to gain a new, updated view of the world. To consider this in non-statistical terms, think for a moment about Columbus stumbling across America.

He found what we now call the West Indies – the name itself a result of his knowledge at the time – and believed he was in India. Despite evidence undermining this, he is said to have gone to his deathbed unconvinced of the alternative, that he had found a new continent. Evidence built up until Amerigo Vespucci made it almost incontrovertible. At this stage in our own knowledge, we don't even consider weighing the evidence for America's existence.

Evidence builds up on top of prior knowledge. What we know now is what we knew before with added new evidence. This form of learning is the key to the Bayesian viewpoint, the immutable relationship between rational belief and evidence as a key feature of science.

Before you feel all at sea, don't worry, we are not so far from shore. We come across the same practical problem with classical statistics: if you end up with 999/1000, do you test it? Nope. So similarly, if so much evidence has built up in 'the prior', to use a Bayesian term, then further calculation becomes unnecessary.

Now we can move to the formal statistics, but as usual without much in the way of formulae. The formal terms for what we have been discussing are the **prior distribution** (or simply 'the prior') and the **posterior distribution** (or posterior). The former is the statistical evidence we have at first. The latter is what we have after the fresh evidence has been calculated:

Prior distribution $\Longrightarrow$ Posterior distribution

At a sophisticated level, it is perfectly possible – especially with an automated system – for the combined knowledge above to become a new prior, meeting new information to create yet another posterior, and so on:

(Prior $\Longrightarrow$ Posterior) $\Longrightarrow$ new Prior

(new Prior $\Longrightarrow$ new Posterior) $\Longrightarrow$ another new Prior

Mathematically, there are computerized steps to be taken once the prior and sample are known. In particular, a **likelihood function** takes your sample data and analyzes the conditional probabilities. The likelihood function is very different from the prior. The Bayesian analysis then finds an intermediary distribution, which fits between these: that is the posterior distribution.

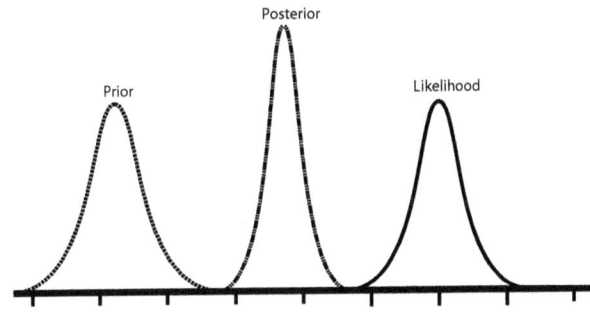

Generally, the preoccupation in Bayesian analysis is between the prior and the posterior, whether or not they are positioned differently.

How are Bayesian statistics used to test hypotheses?

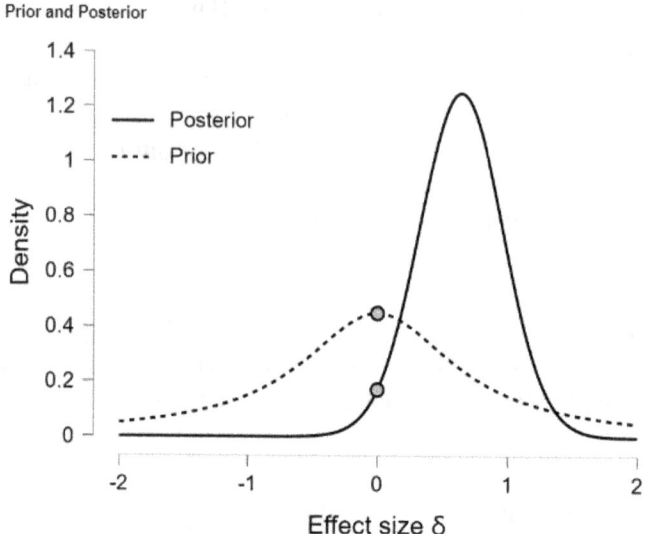

Above, in a chart from a JASP statistical test, we can eyeball the Bayes factor, comparing prior to posterior height at 0. This is somewhat lower than 3, a weak/anecdotal result. (Thank you, E-J Wagenmakers.)

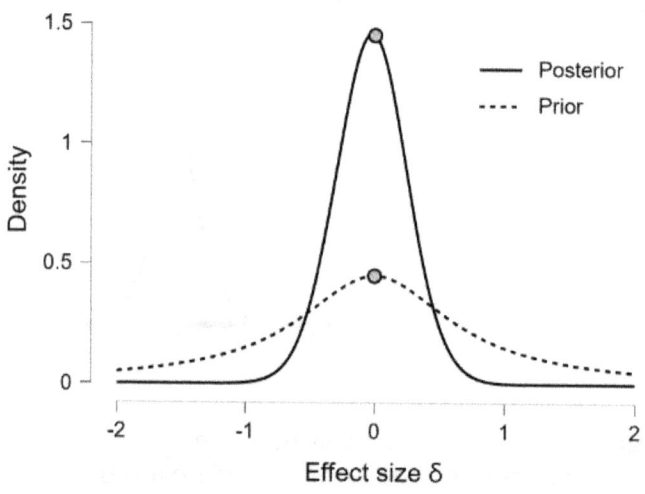

The second chart has the two curves positioned similarly, a clearly insignificant result.

Starting with different priors according to the researcher's current knowledge is almost definitively subjective. Some find this lack of objectivity disturbing; choice of other priors can lead to somewhat different results. On the other hand, it should be noted that the more observations you have, the less important the prior becomes. Also, to my limited mind, the reporting of the priors used should allow replication (just as I have recommended recording your options for other tests). But, see the next section..

And now, even better news!

In practice, people doing what you're doing in this book, testing hypotheses, don't need to devise priors. In general, you will assume a lack of knowledge of the world (with a suspicious similarity to the 'null hypothesis' of classical statistics). So we use **uninformative priors**, the default setting. This can be changed, but if in doubt, leave the default as it is.

For our purposes, the main differences between classical tests and Bayesian tests are the mathematics and their philosophy. On the subject of the mathematics, I will merely note that the recent appearance of computerized Bayesian analyses is no coincidence: Bayesian statistics are simple as fundamental logic, but demand a lot of processing.

Reporting Bayesian results

In classical statistics, we tend to use the p value and the related critical value (such as $p < .05$) to determine whether or not we can reject the null hypothesis, that variance is the result of noise and chance fluctuation. Bayesian statistics are reported more intuitively: evidence is reported *in support* of the alternative hypothesis (and the null hypothesis).

Bayes factors can be placed within reporting bandings, providing measured evidence. The following reporting suggestions are adapted from Jarosz and Wiley (2014):

Statistic		Quantification of evidence	
Bayes Factor (BF_{10})	BF reciprocal (BF_{01})	Jeffreys' interpretation (1961)	Raftery's interpretation (1995)
1 – 3	1 – 0.33	Anecdotal	Weak
3 – 10	0.33 – 0.1	Substantial (substantive, moderate)	Positive
10 – 20	0.1 – .05	Strong	Positive
20 – 30	.05 – .03	Strong	Strong
30 – 100	.03 – .01	Very Strong	Strong
100 – 150	.01 – .0067	Decisive	Strong
> 150	< .0067	Decisive	Very Strong

The Bayes factor, BF_{10}, is evidence in support of the alternative hypothesis. The reciprocal (less mathematically, its converse), is evidence in support of the null hypothesis. These are referred to as BF_{10} and BF_{01} respectively in JASP.

Large numbers in BF_{10} support the **alternative hypothesis**, that the effect being studied is significant; larger numbers in BF_{01} support the null hypothesis.

The lowest reporting level is a good example of how the grid differs from the classical interpretation of statistics in giving us gradations. We are given quite clear guidance: a Bayes factor of between 1 and 3 indicates evidence that is Anecdotal/Weak. In classical statistics, we would instead be scratching our heads with a p value of .052; maybe it is 'insignificant', 'a trend towards significance', or 'a tenuous result'.

Reporting Bayesian results

In practice, results which narrowly attain the classical $p < .05$ critical value typically appear in the 'Substantial' Bayesian banding. This is not always the case, as the two sets of calculations do different things. Lee and Wagenmakers (2013) prefer 'Moderate' to 'Substantial', as the latter seems rather strong (Schonbrodt, 2015). It is of course possible that Jeffreys meant 'substantive', meaning of importance in the real world ('substantial' means of some considerable size). 'Substantive' might fit this category, but as the word is not commonly used and easily confused with 'Substantial', the word actually used in Jeffreys (1961), Lee and Wagenmakers' suggestion of 'Moderate' seems most sensible.

Considering the third category, Rafter considered that there was no evidence to justify Jeffreys' use of 'Strong' here. So he prefers to stay Positive! He selects the fourth category as the threshold for 'Strong'.

Not present is a reporting band for Bayes factors (BF_{10}) of less than 1 and reciprocals (BF_{01}) of greater than 1. Indeed, E-J Wagenmakers, the founder of the JASP statistical package, would prefer there to be a 'noise' category. In counterpoint, his colleague Richard Morey says "a number's a number; why categorize?" (I'll leave you to decide on an answer to that.) Results in this 'noise' area would be dismissed as clearly non-significant.

Statistic		Quantification of evidence
Bayes Factor (BF10)	BF reciprocal (BF01)	
< 1	> 1	Noise
1 – 3	1 – 0.33	Weak
3 – 10	0.33 – 0.1	Moderate
10 – 20	0.1 – .05	Positive
20 – 150	.05 – .0067	Strong
> 150	< .0067	Very strong

While you are free to use the interpretations of Jeffreys (1961) or Raftery (1995), or your own adaptations, I will use this table for the purpose of making reference to Bayesian results during the rest of the book. It is essentially Raftery's interpretation with the use of Moderate rather than Substantial / Substantive, and with Wagenmakers' Noise category.

Other adaptations are possible, and indeed may be desirable. Wei *et al* (2022) are inclined to retitle the BF10 banding of 1 – 3 as 'Not worth more than a bare mention'. Navarro *et al* suggest that this banding could be called 'Negligible', and that what is referred to as Positive evidence could be called 'Weak'; they go on to say that anything between 3 and 20

> should be considered weak or modest evidence at best. But there are no hard and fast rules here: what counts as strong or weak evidence depends entirely on how conservative you are, and upon the standards that your community insists upon before it is willing to label a finding as 'true'.

Context is also considered by Wei *et al*:

> Reporting Bayes factors can be guided by setting customized thresholds according to particular applications. For example, Evett (1991) argued that for forensic evidence alone to be conclusive in a criminal trial, it would require a Bayes factor of at least 1000 rather than the value of 100 suggested by the Jeffreys scale of interpretation.

Regardless of which reporting system is used, I think that issues such as context and effect size should be considered. But one road I do not wish to go down is that of quantifying the relationship between the Bayes factors and the two hypotheses, as in 'this is so many times bigger than that'. To do this we would need to distinguish between the mathematical concepts of likelihood and probability, and the situation pertaining to the prior odds. You can see on the internet people digging around the fine distinctions of such statements. Apart from suggesting that, of the pundits, E-J Wagenmakers is probably the clearest, I would say that for the non-mathematically inclined, that is really not a good place to go!

Three important technical points

Bayesian tests in JASP have similar *assumptions* about data to their Classical / Frequentist counterparts. So the Bayesian equivalent of a classical parametric test requires the same assumptions to be met. (Obviously, the Bayesian counterparts of non-parametric tests such as Wilcoxon and Mann-Whitney will have similarly relaxed requirements.)

The default Prior is the uninformative prior setting of 0.707. I very much recommend that you *use the default* (unless or until you have a more sophisticated understanding of Bayesian analysis).

It should be noted that while JASP's Bayesian tests have familiar names, such as ANOVA, the names have been chosen so that their overall role is easily recognized. They are not completely analogous to the classical tests, being different instruments using different calculations.

Which tests to use, classical or Bayesian?

Often, the tests come out with similar results for the same set of data.
    Many proponents of Bayesian statistics say that you do not have to jump through that rather convoluted hoop of classical statistics, the rejection of the null hypothesis (the negation of a negative). Also, Bayesian tests do not consider the sample against a largely unobserved population (Wagenmakers 2007), but considers the sample in itself, in combination with the known (actually, usually uninformed) current knowledge. This being the case, one should be able to refer to evidence in favor of the alternative hypothesis, the theory of interest to the researcher, or perhaps more profitably, look at different alternative hypotheses, such as X does not equal Y, or more specifically X is bigger than Y (or vice-versa).
    Conversely, this does suggest one really good reason for using classical statistics. If you wish to determine whether or not a hypothesis is true, then the classical/frequentist model makes more sense. Accord-

ing to Popper (1968), in science one can only falsify a theory. But we can ascertain if there is support for the null hypothesis. The p value represents merely the likelihood that the null hypothesis can be upheld, the major focus in classical statistics.

On the other hand, perhaps you have already rejected the null hypothesis and want to determine a model's likely **credibility** given the data. Then the Bayesian approach gives clear(ish) reporting categories. So, for example, if you have a p value of .017, but are not sure of its level of credibility, you may seek a Bayes factor.

A common practical situation in classical statistical testing is the quandary of the uncertainty of a p value close to .05. Do note that its proponent, R. A. Fisher, did not consider this critical value to be an absolute, but a useful place to consider a hypothesis. Let us assume that you have not been dredging (lots of meaningless testing) but have a result which seems a little uncertain: you might wish to complement the classical test with a Bayesian test.

A more philosophically reasonable use of Bayesian techniques than the triangulation just suggested is the comparison of models. A variety of samples may be subjected to examination and it seems reasonable to compare their credibility by reporting their respective Bayes factors.

As this rather contrapuntal discussion suggests, you can to a large extent select the method you prefer. While they are different and provide somewhat different insights, there is no hard and fast rule that says 'use classical for X and Bayesian for Y'. While the null hypothesis may be considered to have its spiritual home in classical statistics, it can be tested using Bayesian methods.

So you can use the methods with which you are most comfortable, or which make the most sense to you. However, you will find different tools available. For example, the classical tests currently provide more measures of effect size, how much of the variance can be attributed to the effect. Also, if you are dealing with non-continuous data such as ordinal data and require an ANOVA, I would stick to good old non-parametric tests such as Friedman and Kruskal-Wallis. You also might want to investigate resampling techniques such as permutation tests and bootstrapping, but that takes us well beyond the remit of this book.

A final consideration is what could be called Stats Wars. The internet is awash with articles telling you how Bayes is the way forward and how 'frequentists' (users of classical statistics) are pseudo-objective, have made lots of terrible errors and are just *so* twentieth century.[*] There are so many of these that I see no point in citing them, but would suggest that their quantity is probably related to the convergence of the wide use of the internet and the speed of twenty-first century computers. For a forceful but well-considered article in favor of Bayesianism and against commonly held frequentist errors, try Wagenmakers (2007).

There are, however, plenty of statisticians who remain frequentists. See, for example, Dennis (1996) and articles by Mayo (2012) and Steinhardt (2014), who suggests that the main reason why so many frequentists have produced 'bad science' is because most science *per se* has been conducted by frequentists.

A short article by Gelman (2011) seeks to decouple myths dividing frequentists and Bayesians. Among other things, he indicates that Bayesian statistics and testing the null hypothesis are perfectly compatible.

Critical thinking, as usual, is more to the point. So, you can use either set of methods or both.

---

[*]The more ebullient proponents of Bayesian statistics consider the 'null or not' way of thinking as being unworldly: "your test suggests that the world won't exist in the morning – I would bet that it will".

A practical example – One Sample T-Test

At the end of the previous chapter, we used the Classical version of the single sample *t* test. In the example we used, the results were *p* = 0.26 if a one-tailed hypothesis were adopted and .052 for two-tailed (assuming a critical value of *p* < .05). * Let's now see what the Bayesian equivalent makes of the same data.

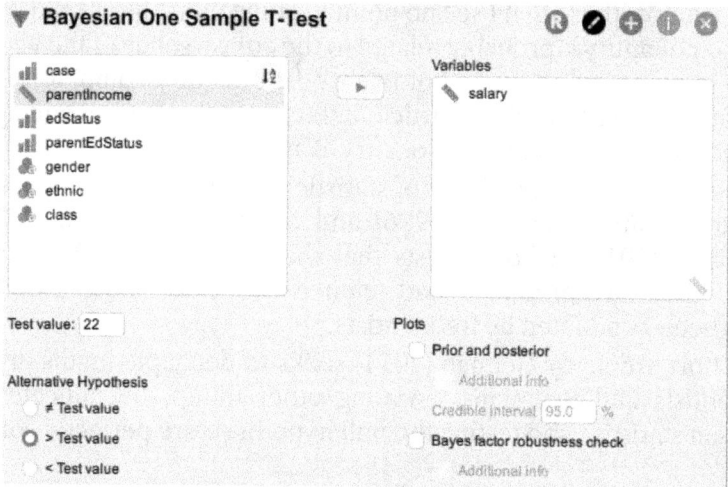

Again opening **Basic.csv**, choose Bayesian / One-Sample T-Test, again using the Test value of 22 and the one-tailed hypothesis of '> Test value'.

Bayesian One Sample T-Test

	BF_{-0}	error %
salary	3.275	~ 0.008

Note. For all tests, the alternative hypothesis specifies that the population mean is greater than 22.

---

*As noted in the previous chapter, in real life you should not adopt two incompatible hypotheses at once.

A practical example – One Sample T-Test

Statistic		Quantification of evidence
Bayes Factor (BF10)	BF reciprocal (BF01)	
< 1	> 1	Noise
1 – 3	1 – 0.33	Weak
3 – 10	0.33 – 0.1	Moderate
10 – 20	0.1 – .05	Positive
20 – 150	.05 – .0067	Strong
> 150	< .0067	Very strong

Consulting our reporting guide, a Bayes factor of a little over 3 means that it just about attains the 'Moderate' banding in favour of the alternative hypothesis. (The reporting bands indicate the credibility of the evidence.)

What if we had opted for the two-tailed hypothesis?

Bayesian One Sample T-Test

Test value: 22

Alternative Hypothesis
○ ≠ Test value
○ > Test value
○ < Test value

	BF_{10}	error %
salary	1.699	4.763×10^{-6}

Note. For all tests, the alternative hypothesis specifies that the population mean differs from 22.

A Bayes factor of 1.699 puts the result well down into the Weak banding. This interpretation suggests that the alternative hypothesis still retains a little credibility. Bayesian reporting gives us a more measured outcome than the significance-based results of classical reporting.

Chapter 6 – Tests of differences

Design considerations for the analysis of differences

We need to consider research design every time we think about posing a question. This also applies to choosing the most appropriate test. Of particular concern is the issue of Related versus Unrelated design. You will come across other terms in JASP and the statistical literature:

Related design	Unrelated design
Same subjects/participants	Different subjects/participants
Paired design	Independent design
Within–subjects design	Between–subjects design
Paired	Unpaired
Matched	Unmatched
Repeated measures (the same over time)	
Panel data (the same people over time, but a term used in business statistics / econometrics)	

As an example, let us suppose that we were planning a study of social exclusion in a region's schools. Before the study begins in earnest, we might want to find out if pupils' attitudes to ethnic and social differences tend to be different at the start and the end of the school week, in relation to the contact hypothesis.

We may choose a **related design**, studying the same people. We could have all pupils tested on Mondays and Fridays, as perhaps feelings are influenced by the experiencing of a school week. A related design is generally preferred because it eradicates individual differences, but it may not always be feasible, if we need individuals to be fundamentally different, as in studies of gender or status differences, or perhaps it is impractical in the particular context. In this example, we run the risk of people becoming bored with the tests or some form of experimental contamination in the intervening period. With two conditions, we would consider using the paired samples t test or the non-parametric Wilcoxon signed rank test.

In an **unrelated design**, in which we study different people, we may decide to have half of the schools tested on a Monday and the others on a Friday. Testing different people rids us of carry-over effects, such as students becoming bored with two tests in a week. However, such a design introduces the problem of individual differences, in this case applying to both the students and potentially different school environments. At this stage, for two conditions, we would consider using either the independent samples t test or its non-parametric equivalent, the Mann-Whitney U test. These tests of independent samples allow to some extent for individual differences.

Some researchers like to attempt **pairing**, also known as **matching**. Although the individual participants are not the same, they are grouped for characteristics which are deemed relevant to the study. In this example, we may decide to pair up schools from similar social milieu, and maybe even students from the same ethnicity. In such a situation, this could be seen as a related design, so the use of tests for same subjects (paired samples t or Wilcoxon) would be deemed appropriate.

In addition to the question of related and unrelated design, we also need to consider if we are using *more than two conditions*. So far we have considered only two conditions. Let us say that we decide to run our tests on a Wednesday as well as on Monday and Friday, giving 3 conditions, or perhaps every weekday, making 5 conditions.

Design considerations for the analysis of differences

Another example would be if we were to try out three different social interventions (Intervention 1, Intervention 2, Intervention 3), or two interventions and a control condition. For more than two conditions, we would consider using, for the related design, either the Repeated Measures ANOVA or the non-parametric Friedman test; for the unrelated design, we would consider ANOVA or the non-parametric Kruskal-Wallis test.

Yet another consideration, which helps us to narrow down to the test of choice, is *the type of data* being examined. In this chapter, data must at least be comparable numerically (categorical, or 'nominal' data appears in Chapter 8, which covers frequencies of observation). If the data is continuous, it is a candidate for parametric tests, such as the T Test and ANOVA (Analysis of Variance); as mentioned previously, but we also need to consider the assumptions of normality of distribution and homogeneity of variance covered in Chapter 3. If these assumptions are not met, or the data is ordinal, then we should abandon parametric tests and should consider non-parametric tests, including the Wilcoxon, Mann-Whitney, Friedman and Kruskal-Wallis.

Some research design terminology applied

When we deliberately set up an intervention, including the allocation of cases to different conditions, we are running an experiment (with or without a laboratory). We manipulate variables, things which are changeable (variable). We actively manipulate an independent variable, the effect of the time of the week in the recent example; it could be, however, the type of social intervention, or a single intervention over time. We observe the effect of manipulating the independent variable by looking at changes in the dependent variable, the measure; this could be, for example, test results, numbers of social interactions or income as a result of education.

The independent variable is varied by the experimenter. The dependent variable, the measure, is data which is dependent on such variations. The terms independent variable and dependent variable

Design considerations for the analysis of differences

should, strictly speaking, only be used when referring to experiments, but in practice they are used much more widely.

Very many 'real world' analyses of differences are quasi–experimental. We do not manipulate variables ourselves, but use records or observations without allocating groups in any organized way. If we ran a quasi–experimental version of our study, the effect of which part of the week would be called a predictor (instead of independent variable) and the test result would be the criterion (instead of dependent variable). The criteria for measurement can be test scores, income, reduction in incidents, attitudes to social phenomena and so on.

While the terms predictor and criterion should be used in non––experimental research, they are often interchangeable with independent and dependent variables in the literature. Both terms are used in the following example (ignore the results; there are too few observations).

	Independent Variable / Predictor: Part of the week	
	Condition 1: Monday	Condition 2: Friday
Dependent Variable / Criterion: Attitude scale	40	50
	30	40
	45	45
	60	50
	45	43

To practice using the statistical tests in this chapter, we use the variables held in the file **Differences.csv**, although the data is also presented on the page so that you can create your own files should you wish.

Tests for same subjects

Paired Samples *t* test: a parametric test for two conditions, same subjects

We are interested in the effects of alcohol on memory and ask university students to sleep over at a laboratory on two separate occasions. On any given night, half went to bed after drinking two units of alcohol, while half went to bed with four units. In both cases, they were told a not particularly interesting story before going to bed and were woken two hours later to recall the story. If there was a noticeable difference between the scores, we want to know if such a difference was statistically significant. As each individual underwent both conditions at some point, this is a related design.

		Predictor: Alcohol consumption	
	Experimental participants	Condition 1: Alcohol – 4 units	Condition 2: Alcohol – 2 units
Criterion:	1	52	60
	2	53	34
	3	47	38
	4	40	52
Recall score (as a percentage)	5	48	54
	6	45	55
	7	52	36
	8	47	48
	9	51	44
	10	38	56

Open JASP. Using the tab on the far left, find the **Differences.csv** file.

Tests for same subjects

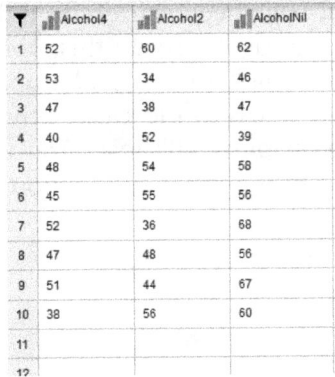

The two relevant variables are Alcohol4 and Alcohol2. The next one, AlcoholNil, will extend our example to three variables when we use the Repeated Measures ANOVA a little later.

Press the T-Tests tab at the top and choose Classical / Paired Samples T-Test from the menu.

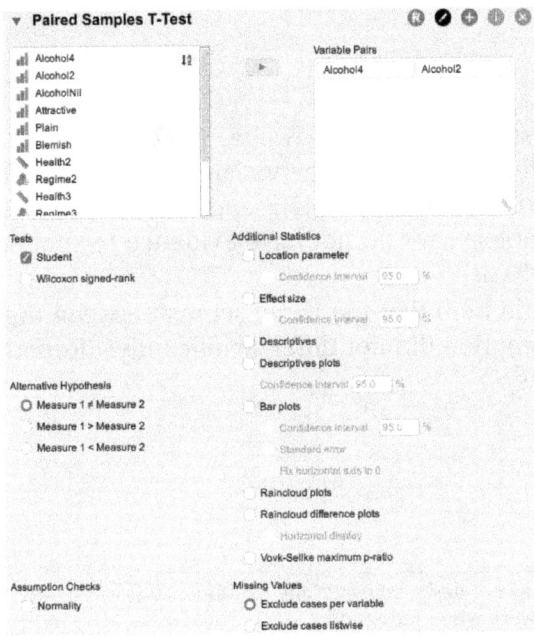

Tests for same subjects

Alcohol4 and Alcohol2 have been shifted from the list of variables on the left to the box on the right. To do this, you can use the transfer (arrow) button, or double-click, or drag and drop. We retain the default option, Student's t test. * Go to the Descriptives menu for the most reliable version of the Shapiro-Wilk test. You will find that Shapiro-Wilk is non-significant, so a normal distribution is assumed, and the t test can be used.

Paired Samples T-Test

Measure 1		Measure 2	t	df	p
Alcohol4	-	Alcohol2	−0.101	9	0.922

Note. Student's t-test.

The t test result has a huge p value, 0.922, so there is clearly no significant difference between the variables (if you use the Descriptives option, you will see means of 47.3 and 47.7 respectively). Using the correct terminology, we do not have evidence to support rejection of the null hypothesis.

We conclude here that the recall scores between the two levels of alcohol consumption did not differ significantly. (Reporting note: you cite t as well as p.)

---

*Sorry, this was not designed especially for students; it was named after a chemist at the Guinness brewery who used Student as a pen-name.

Tests for same subjects

Bayesian equivalent

Statistic		Quantification of evidence
Bayes Factor (BF10)	BF reciprocal (BF01)	
< 1	> 1	Noise
1 – 3	1 – 0.33	Weak
3 – 10	0.33 – 0.1	Moderate
10 – 20	0.1 – .05	Positive
20 – 150	.05 – .0067	Strong
> 150	< .0067	Very strong

This is a reminder of a suggested guide to Bayesian hypothesis reporting.

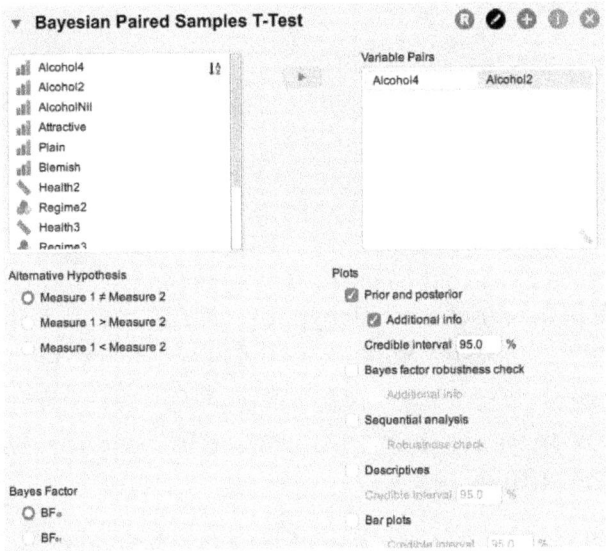

We choose Bayesian / Paired Samples T-Test. While the underlying algorithms are different, data entry for a JASP Bayesian equivalent is the same as for its classical tests.

Tests for same subjects

Bayesian Paired Samples T-Test

Measure 1		Measure 2	BF_{10}	error %
Alcohol4	-	Alcohol2	0.310	0.005

A tiny Bayes factor indicates no support for the alternative hypothesis. Any 'effect' is little more than noise; there is agreement with the classical ('frequentist') test.

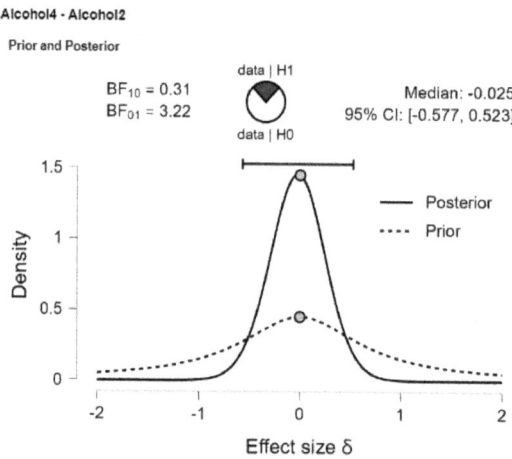

The posterior is close to the centre. The pizza shows the predominance of evidence in favour of the null hypothesis, *H0*; the dark section (dark red in real life) represents support for the alternative hypothesis, *H1*.

The Wilcoxon test: a non-parametric test for two conditions, same subjects

In an experiment to test the effect of facial attractiveness on observers' emotional responses, each participant sees a photograph of a fictional team of welfare workers with conventional good looks (Condition 1) and another team with plain looks (Condition 2). As the same participant undergoes both conditions, this is a related design. A 5 point scale rates the teams' perceived helpfulness, higher ratings denoting greater helpfulness.

Two methodological points should be considered. In this example, the scale was not calibrated before use, and so cannot be viewed as continuous data. Without calibration, a non-parametric test is preferred. Also, given a small data set, we settle for a critical value of $p < .05$. As we are not really sure which scheme would be more effective, we opt for the more rigorous two-tailed hypothesis.

	observer	Predictor: attractiveness	
		Condition 1: Attractive	Condition 2: Plain
Criterion:	1	4	5
	2	3	3
Helpfulness	3	2	4
(1 to 5 Likert scale)	4	4	5
	5	3	5
	6	4	2
	7	3	3
	8	5	4
	9	3	5
	10	4	5
	11	3	5
	12	2	4
	13	2	5

We use the T-Test tab and again choose the Classical / Paired Samples T-Test option. As in the previous example, transfer the relevant variables to the right, here the Attractive and Plain variables. Under the Tests options, select 'Wilcoxon signed-rank', the non-parametric test we require, deselecting the Student's test, which is the parametric test we used previously. From the Additional Statistics section, choose Effect size; also select Descriptives.

Tests for same subjects

Paired Samples T-Test

Paired Samples T-Test

Measure 1		Measure 2	W	z	df	p	Rank-Biserial Correlation
Attractive	-	Plain	56.000	2.045		0.041	0.697

Note. Wilcoxon signed-rank test.

Descriptives

Descriptives

	N	Mean	SD	SE	Coefficient of variation
Attractive	13	4.231	1.013	0.281	0.239
Plain	13	3.231	0.927	0.257	0.287

We are interested in the Wilcoxon's W statistic (ignore the T-test header). First, we look at the p value: .041 comes within the $p < .05$ critical value.

If we had expected beforehand, theoretically and without eyeballing the data, that Attractive was going to be rated higher than Plain, we could have selected a one-tailed test. This would have been achieved by changing the Hypothesis to Measure 1 > Measure 2. The resulting p value for the Wilcoxon test, .02, would be smaller.

A thinking point: If you look at the t test result for this pairing, you will find that the Wilcoxon is more conservative. This is intentional, given the roughness of the data. However, if you use data suitable for parametric tests, you will often find that both the Wilcoxon and t tests give the same results. I think this important when considering the claims of those who advocate the use of parametric tests in all circumstances. They are 'more powerful', we are told, and non-parametric tests are old-fashioned, always a reason for throwing away something that works!

As we have evidence to support rejection of the null hypothesis, we are interested in the effect size (the option is to be found under 'Additional Statistics'). Unlike 'significance', this measures the magnitude of the difference between the variables. Effect sizes are to some extent dependent on context, but generally for t tests, from 0.2 to 0.3 is considered small, 0.8 upwards is large (this statistic can exceed 1.0);

in-between values are medium. In this example the value is 0.697, a medium-sized effect.

JASP offers further descriptive statistics within the Descriptives menu, including the median, which I prefer to the mean as the central tendency statistic for non-parametric tests. However, you will find when handling Likert scales that the mean gives superior differentiation when handling results that are less clear-cut than those in our current example. When examining a pair of effects, one thing that you can do is to calculate the *mean difference*, or *median difference*, by subtracting one mean (or median) from the other.

Bayesian equivalent

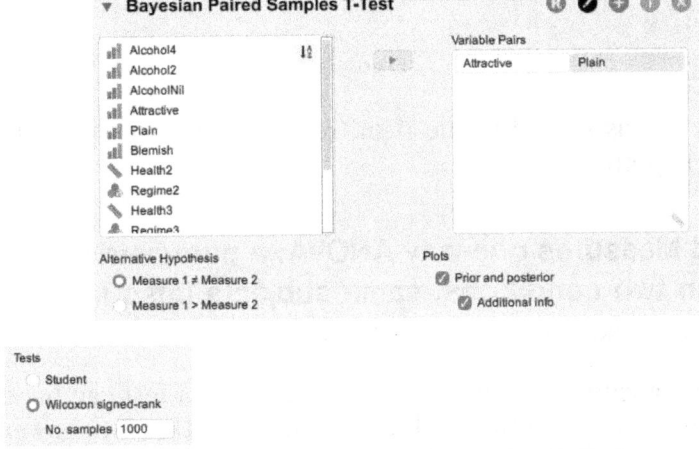

Further down the dialog, we switch from the default parametric test ('Student') to its non-parametric equivalent.

Bayesian Wilcoxon Signed-Rank Test

Measure 1		Measure 2	BF_{10}	W	Rhat
Attractive	-	Plain	3.129	56.000	1.001

We have agreement with the classical test; a Bayes factor above 3 takes the result into the Moderate banding. If we had followed the one-tailed

Alternative Hypothesis, using Measure 1 > Measure 2, we get a result just above 6, within the same banding of credibility. *

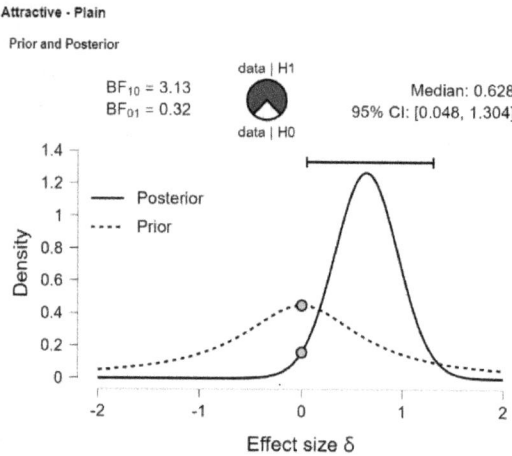

The posterior has moved to the right, reflecting the direction of the alternative hypothesis.

Repeated Measures one-way ANOVA: a parametric test for more than two conditions, same subjects (also known as the Within-subjects one-way ANOVA)

As the title 'repeated measures' suggests, this type of test can be used for the analysis of time series, including panel data. Do note, however, that there is quite a lot to know about time series and that some reading around the subject is advisable. In this example, the primary focus is on different conditions.

Here, we use our earlier *t* test example, where we contrasted the results of recall with differing levels of alcohol consumption. Here

---

*You may see some differences in the results each time the Bayesian Wilcoxon is used. This is because of its underlying algorithm, which you will notice also takes some time to process. Greater stability in Bayes factor results may be achieved by raising the iterations (No. samples) from 1000 to 10,000 (Goss-Sampson, 2020).

we decide to include a control condition when the students went to bed without any alcohol (as with the previous example, the results are fictitious).

Predictor: Alcohol levels or no alcohol
Condition 1: 4 units **Condition 2:** 2 units **Condition 3:** No alcohol

The additional results are: 62, 46, 47, 39, 58, 56, 68, 56, 67, 60, which are to be found in the variable AlcoholNil from the Differences.csv data file.

First we look at the descriptive data. *You should always do EDA (exploratory data analysis) before running an ANOVA.* We need to check the assumptions of normal distribution and that the data is suitably measurable.

For the Shapiro-Wilk test of normality, I suggest for consistency that you use the Descriptives menu tab. In this example, you will find that all of these are non-significant. Additionally, the Statistics panel has a range of Distribution statistics including Skewness and Kurtosis. While you are there, you can also check that your data is legitimate by looking at the Dispersion values; these will help you to check for peculiar or illegitimate data.

However, for ANOVA, there are more assumptions to meet. To see if the data meets these assumptions, we need to set up the ANOVA test itself. Press the ANOVA tab and select Classical / Repeated Measures ANOVA from the menu, revealing this mysterious-looking interface:

Tests for same subjects

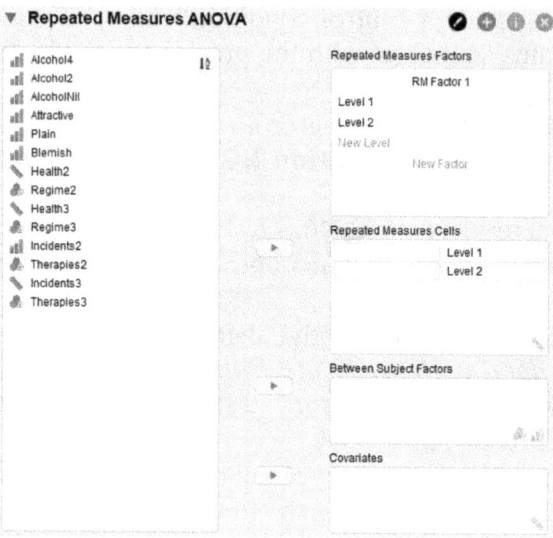

This is a little more difficult to set up than the *t* test, but compared to doing it with SPSS, it's a breeze.

The first thing we should do is to replace RM Factor 1 with a more meaningful name; using the title of the predictor / independent variable would make sense. In the one-way ANOVA, this will assist your reader's understanding of the output; when you get to use a two-way (or three-way) ANOVA, you will need it to assist your own understanding!

Then replace the levels (another term for conditions) starting with Level 1, by typing in titles for the variables. If you use post hoc tests, or the descriptive plot, these will appear. You will probably want to provide somewhat more detailed names than those given to the variables, but if they are too long, they may look confusing on some of the output.

You only need RM Factor 1 for one-way ANOVA; further factors come into play when you use a factorial ANOVA (to be covered in Chapter 11). Similarly, you only need as many levels as the conditions you are analyzing (which reminds me: if you use too many variables, you are increasing the danger of a Type 1 error, thinking that you have a significant finding when you don't).

At this stage, the top left of the Repeated Measures ANOVA interface should look like this:

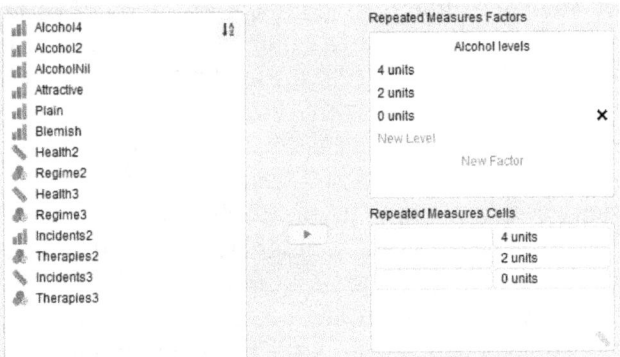

The titles of the levels typed in underneath the factor name have automatically been transferred to the right-hand side of the Repeated Measures Cells box. Then comes the easy bit: transferring the variables themselves to the left-hand cells. Do be careful as your data sets get more complicated: the transferred variables on the left must match the levels on the right.

The variables are transferred by using the arrow or dragging. So you should then have this:

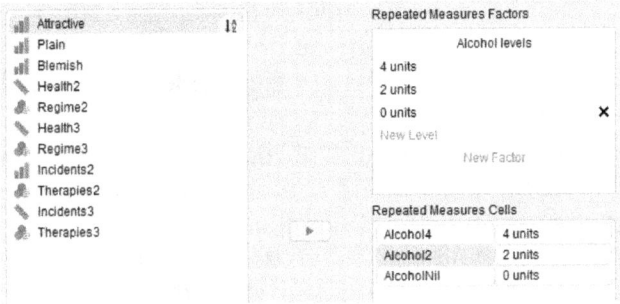

You will notice that I have not shown the lower boxes. The Between Subject Factors box only comes into play when we use a mixed design ANOVA, combining within- and between-subjects factors (Chapter 11).

Tests for same subjects

The Covariates box is for variables of no particular interest but which might influence interactions.

Although statistics immediately appear on the ANOVA table on the display on the right, you need to ignore them for the moment. The results of an ANOVA are likely to be misleading if the data does not meet certain assumptions. As well as the usual assumptions for parametric tests in general, the repeated measures ANOVA has an assumption of sphericity (equality of variances between each pair of levels). Problems with sphericity do not mean that you won't be able to use the ANOVA, but you would need to use a non-standard read-out. So before looking at the results, you need to open the Assumption Checks section and select Sphericity tests. At the moment, we are only interested in Mauchly's W. If this is significant at the level of $p < .05$, then the assumption is violated.

Assumption Checks

Test of Sphericity

	Mauchly's W	Approx. X^2	df	p-value	Greenhouse-Geisser ε	Huynh-Feldt ε
Alcohol levels	0.846	1.342	2	0.511	0.866	1.000

In this case, however, a p value of .511 is healthily non–significant, so we can ignore sphericity as a problem. The figures to the right are only of relevance if p is significant. If you do find yourself with a sphericity problem, refer to Chapter 11, where the problem arises in the ANOVA mixed design worked example.

Before considering the results of the ANOVA itself, let us quickly consider the concepts which are central to analysis of variance. The **variance** is the variability of data around the central tendency (usually the mean); if observations tend to vary a lot from the mean, the variance is large, and vice-versa. **Analysis of variance** calculates how much variance comes from independent variables and how much is due to error (error variance). The calculation, the variance divided by the error, is the F ratio, referred to in the output as 'F'. Essentially, the bigger the F ratio, the more likely it is that the effect is a significant one.

Tests for same subjects

Within Subjects Effects							
Cases	Sum of Squares	df	Mean Square	F	p	η^2
Alcohol levels	471.200	2	235.600	3.637	0.047	0.288
Residuals	1166.133	18	64.785			

Note. Type III Sum of Squares

The ANOVA table shows an F ratio of 3.64 and a p value of 0.047. Assuming that we accept a critical value of $p < .05$, our effect may be considered to be a significant one. There is evidence to support the rejection of the null hypothesis of no mean differences across the three conditions.

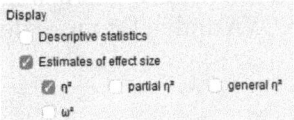

We can also look at the effect size. One rule of thumb for ANOVA effect size is that 0.01 is small, 0.06 is medium and 0.14 is large. The default Eta squared indicates a large effect of .288. Partial Eta squared is preferred for factorial ANOVA.

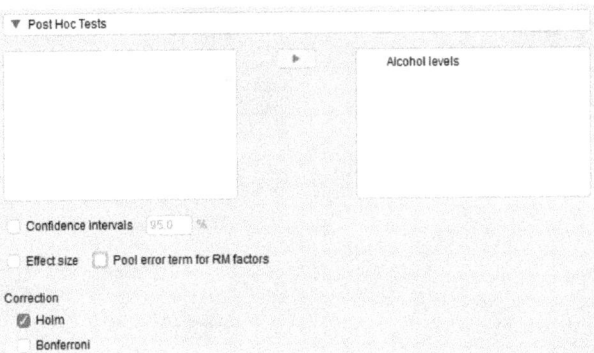

If you wish to look at individual pairings, open the Post Hoc Tests section and transfer the relevant variable to the right. The Bonferroni correction is often the default in statistical literature, but it is nowadays considered too severe in most contexts. (I prefer not to use the default 'Pool error term' option.)

Chapter 6 – Tests of differences

Tests for same subjects

Post Hoc Tests

Post Hoc Comparisons - Alcohol levels

		Mean Difference	SE	t	p_{holm}
4 units	2 units	-0.400	3.964	-0.101	0.922
	0 units	-8.600	2.806	-3.065	0.040
2 units	0 units	-8.200	3.910	-2.097	0.131

Note. P-value adjusted for comparing a family of 3

There is reason to believe that one of the pairings is of particular importance. These tests will be discussed a little more when we use independent samples (between-subjects) ANOVA and in far more detail towards the end of Chapter 11.

Bayesian equivalent

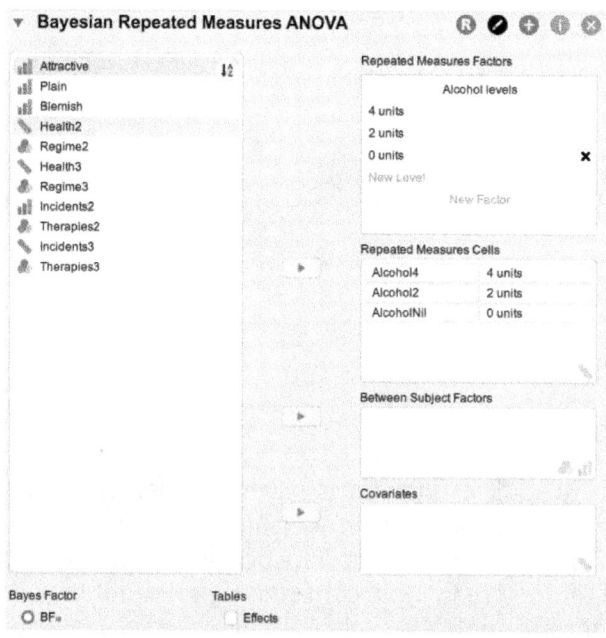

66   The fast guide to statistical testing with JASP

Tests for same subjects

Order		Limit No. Models Shown		
☐ Compare to best model		○ No		
○ Compare to null model		○ Yes, show best 10		

Model Comparison

Models	P(M)	P(M|data)	BF$_M$	BF$_{10}$	error %
Null model (incl. subject and random slopes)	0.500	0.253	0.339	1.000	
Alcohol levels	0.500	0.747	2.951	2.951	0.480

Note. All models include subject, and random slopes for all repeated measures factors.

The Bayesian version of repeated measures ANOVA gives an alternative hypothesis result within the Weak banding, below 3. It should be noted, however, that "the Bayes factor really is a continuous method of evidence" (Goss-Sampson, 2020).

Post Hoc Comparisons - Alcohol levels

		Prior Odds	Posterior Odds	BF$_{10, U}$	error %
4 units	2 units	0.587	0.182	0.310	0.005
	0 units	0.587	2.874	4.892	7.112×10^{-6}
2 units	0 units	0.587	0.846	1.441	1.064×10^{-5}

There is agreement between the two approaches about the 4 units–0 units pairing (moderate). However, whereas the 2 units–0 units pairing is seen as insignificant according to the frequentist approach, the Bayesian test indicates a weak effect (between 1 and 3).

Friedman: a non-parametric test for more than two conditions, same subjects

If the data fails to meet the assumptions of continuous data and normality, use the Friedman test instead of Repeated Measures ANOVA.

In addition to the two conditions examined by the Wilcoxon test, photographs of plain and attractive teams judged for helpfulness, we introduce photographs of a team with mild facial differences such as acne or a squint. As the scale has not been calibrated, a non–parametric test is preferred.

Tests for same subjects

Should you wish to input your own data, the new condition contains the following data: 5 3 2 2 2   2 1 3 2 4   2 3 1, or use **Differences.csv**. Use the same set-up as with Repeated measures ANOVA: press the ANOVA tab and select Classical / Repeated Measures ANOVA from the menu.

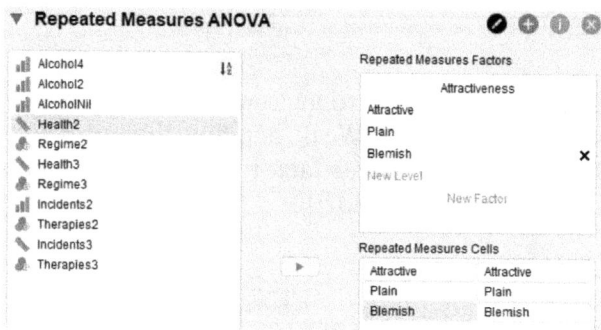

Then we open the Nonparametrics section:

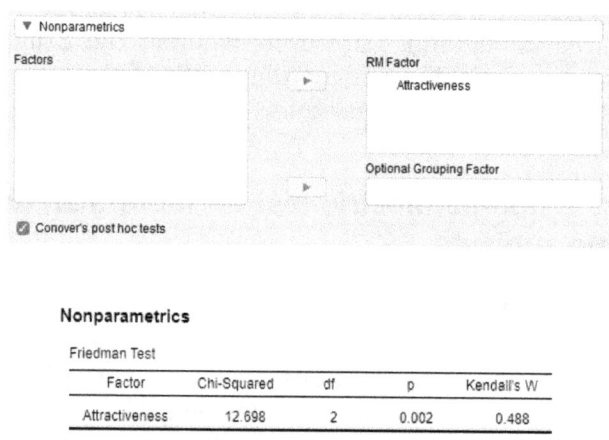

Nonparametrics

Friedman Test

Factor	Chi-Squared	df	p	Kendall's W
Attractiveness	12.698	2	0.002	0.488

There is evidence to support the rejection of the null hypothesis.

Tests for same subjects

Conover Test

Conover's Post Hoc Comparisons - Attractiveness

		T-Stat	df	W_i	W_j	p	p_{bonf}	p_{holm}
Attractive	Plain	1.958	24	34.500	25.500	0.062	0.186	0.124
	Blemish	3.590	24	34.500	18.000	0.001	0.004	0.004
Plain	Blemish	1.632	24	25.500	18.000	0.116	0.347	0.124

Note. Grouped by subject.

The post-hoc tests show a very low *p* value for Attractive-Blemish. The effect is very clear when you examine the Descriptives Plots:

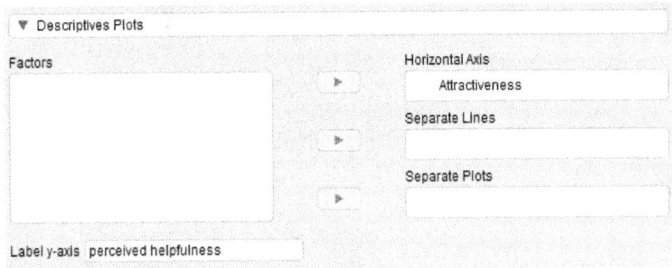

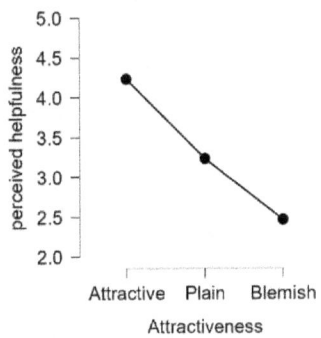

Descriptives plots

---

*At the time of writing, there is no Bayesian equivalent of the Friedman test.

Tests for different subjects

Independent Samples T-Test: a parametric test for two conditions, different subjects

We are interested in the effects of different types of criminal sentence upon mental health. A personality test has been created using a composite of a locus of control scale and other measures; higher scores represent relatively positive health. The independent variable is sentence type: on probation or in a medium security prison.

Sentenced person	Criterion / dependent variable	Predictor / grouping
	Score on test Name: **Health2**	Sentencing regime Name: **Regime2**
1	80	probation
2	68	probation
3	77	probation
4	78	probation
5	85	probation
6	82	probation
7	79	probation
8	76	probation
9	77	probation
10	83	probation
11	84	probation
12	82	probation
13	81	probation
14	80	probation
15	56	medium
16	69	medium
17	73	medium
18	70	medium
19	61	medium
20	65	medium
21	59	medium
22	60	medium
23	53	medium
24	61	medium
25	62	medium
26	71	medium

Data entry up until now has been for same-subjects design, with one column of values next to another. However, as previously noted in Chapter 1, data entry for between-subjects design is a little less straightforward: as an observation or participant needs its own row, different conditions need a demarcation method. Each case has to have its own grouping variable, whether using names (such as medium) or numbers. So cases 1 to 14 are given the probation title, while cases 15 to 26 are medium.

Tests for different subjects

In the **Differences.csv** file, as we have a few simple projects together, only the active variables are included, here the numeric variable Health2 and the grouping variable Regime2. In real life research, you would save identification variables (here, Sentenced person). In more complex research, case numbers are particularly helpful in error-checking; when you are rushed and/or tired, errors do creep in.

Notice that in this study, there are different numbers of cases in the different conditions, 14 on probation and 12 in medium security prison. Only same subject studies are required, quite logically, to have the same numbers in each condition.

To run the test, press the T-Tests tab and select Classical / Independent Samples T-Test from the menu.

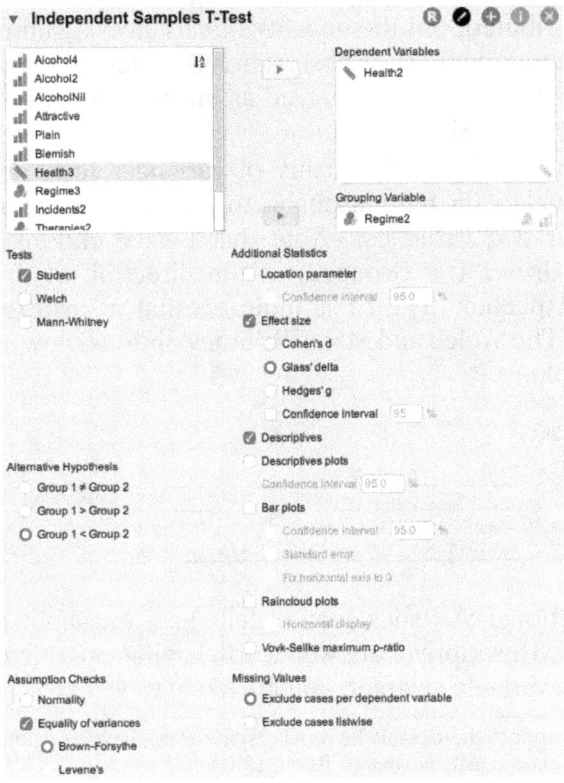

Tests for different subjects

The Regime2 variable has two conditions, probation and medium, and is named to distinguish it from another variable to be used later which has three conditions. It has been transferred to the Grouping Variable box. The corresponding Health2 goes into the Dependent Variables box.

You will find that you have different preferences as you get used to the software. Let us consider the choices taken above. For normality, I would go to the Descriptives tab and use Shapiro-Wilk test there. I would always use the Assumption Checks for Equality of variances (also known as homoscedasticity) * The default t test, Student's, is a parametric test; the data should meet both assumptions as well as being continuous. If the data is clearly not continuous, or the normality assumption has not been met, then you definitely want Mann-Whitney U as a non-parametric test. If the data is continuous and generally normally distributed, but the equality of variances assumption has not been met, or the numbers of observations in each group differ greatly, then Welch's test is to be preferred as an intermediate test between Student's t test and Mann-Whitney.

As both normality and Equality of variances tests show p values greater or equal to .05, it is reasonable to stick with the default Student's t test for our two variables. Note that I have chosen a one-tailed hypothesis (Group 1 < Group 2) as the direction of the effect was very much expected. The t test indicates that we can reject the null hypothesis. (The Welch and Mann-Whitney options show similar results in this example.)

Descriptives

Group Descriptives

	Group	N	Mean	SD	SE	Coefficient of variation
Health2	medium	12	63.333	6.286	1.815	0.099
	probation	14	79.429	4.274	1.142	0.054

The Additional Statistics section tells us a lot about the data. In particular, the Descriptives are well worth having: apart from reminding you of which variable is larger, you are likely to want to cite the means

---

*In general, report the default Brown-Forsythe test. Levene is only slightly more powerful if your data is truly normal (O'Brien, 1981).

and standard deviations (SD) in your reports. The SD statistics also help to decide on an appropriate effect size statistic: Hedge's g is best for significantly different sample sizes, also for when there are fewer than 20 cases. Otherwise, use Cohen's d when standard deviations are similar, or Glass's when SDs are clearly different. Here, I chose Glass's Delta.

Independent Samples T-Test

	t	df	p	Glass' delta	SE Glass' delta
Health2	-7.731	24	< .001	-3.766	0.813

Note. For all tests, the alternative hypothesis specifies that group *medium* is less than group *probation*.
Note. Glass' delta uses the standard deviation of group probation of variable Regime2.
Note. Student's t-test.

The larger the value of t, the easier it is to reject the null hypothesis.

For effect sizes for t tests, from 0.2 to 0.3 is considered small; 0.8 upwards is large (these statistics can exceed 1.0); in-between values are medium. The current effect size of 3.766 is huge. Oh yes, and we have evidence to support the rejection of the null hypothesis; the null would state no difference between the two conditions.

Statistics that you may find useful, using 'Location parameter', are the 'Mean difference' and 'Confidence interval'. The first is a measurement of difference between the two samples, by subtracting one mean from the other; like the mean and the median, this is a 'point estimate', a single statistic which acts as a representative value but has some limitations. The confidence interval is intended to show a range of likely values around the point estimate, citing two figures, the lower and upper confidence limits. Note that if your Hypothesis setting is one-tailed (in this case, Group 1 < Group 2) then one limit will be infinity (∞).

		95% CI for Mean Difference	
Mean Difference	SE Difference	Lower	Upper
-16.095	2.082	-20.392	-11.799

To demonstrate both confidence limits, I have moved to the two-tailed hypothesis option.

Tests for different subjects

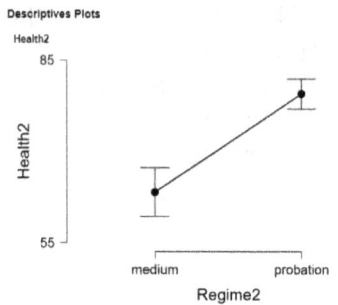

The chart created by the Descriptives plots option shows the clear disparity between the two conditions. The bars represent the confidence intervals, showing a range of likely values. Do note that the confidence intervals do not show the whole range of possible values - note the 95% cited – and they are not without their critics (Morey *et al* 2016).

Our (fictional) evidence indicates that inmates of medium security prisons have worse mental health than those on probation, at least according to these tests. It might be useful to consider other factors: do different sentences reflect differing levels of psychological problems among the convicted, and are there gender or ethnicity differences?

Bayesian equivalent

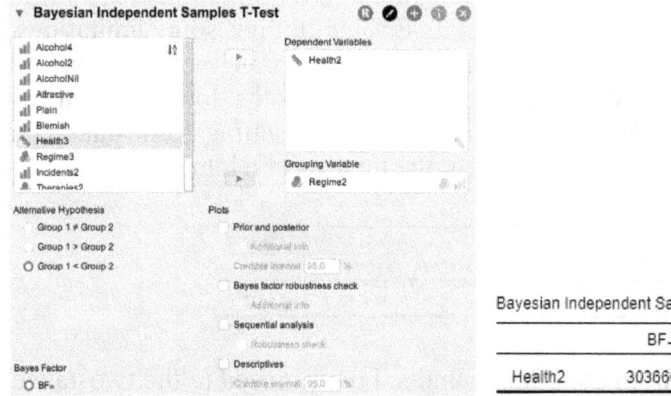

The effect's credibility is adjudged to be very strong indeed.

Tests for different subjects

The Mann-Whitney test: a non-parametric test for two conditions, different subjects

We are interested in the effects of two different psychotherapies on the number of late night screaming outbursts of elderly men with Alzheimer's disease. As each therapy is only available in some parts of the country, patients cannot be offered more than one, so an unrelated design is appropriate, with different patients in each condition.

	Predictor: Type of psychotherapy			
Criterion:	**Therapy A**		**Therapy B**	
Number of reported screaming incidents per person in a week.	Patient 1	5	Patient 11	6
	Patient 2	4	Patient 12	15
	Patient 3	16	Patient 13	4
	Patient 4	6	Patient 14	4
	Patient 5	7	Patient 15	6
	Patient 6	22	Patient 16	7
	Patient 7	8	Patient 17	16
	Patient 8	9	Patient 18	7
	Patient 9	9	Patient 19	5
	Patient 10	8	Patient 20	4

We convert the data, as necessary for a between-subjects design. These are the necessary variables (omitting the case numbers used in real life).

Therapies2	Incidents2
A	5
A	4
A	16
A	6
A	7
A	22
A	8
A	9
A	9
A	8
B	6
B	15
B	4
B	4
B	6
B	7
B	16
B	7
B	5
B	4

Chapter 6 – Tests of differences

Tests for different subjects

As in the previous worked example, press the T-Tests tab and select Classical / Independent Samples T-Test from the menu. Transfer the relevant variables to the right: Incidents2 goes into the Dependent Variables box, with Therapies2 as the Grouping Variable.

The test for normality of the distribution is significant, indicating a clear violation of the assumption, so we require a non-parametric test. Under the Tests option, choose Mann-Whitney, and deselect Student. (Choose Welch's *t* test if you have normality, but Levene's test is significant or there is a big difference between the numbers in each group.)

As there is no expected direction, choose the two-tailed hypothesis. Choose Descriptives, as descriptive statistics are usually useful.

Independent Samples T-Test

	W	df	p
Incidents2	68.500		0.170

Note: Mann-Whitney U test.

Descriptives

Group Descriptives

	Group	N	Mean	SD	SE	Coefficient of variation
Incidents2	A	10	9.400	5.502	1.740	0.585
	B	10	7.400	4.427	1.400	0.598

A p value of 0.17 resulting from the Mann-Whitney means that we cannot reject the null hypothesis, that the two groups are not particularly different.

The descriptives table shows a difference in means. You might be tempted to look at the data derived from a one-tailed hypothesis, Group 1 > Group 2. If you did, you would see p = .085 for the Mann-Whitney test, which could be seen as a trend towards significance. However, unless you have a theory or rationale supporting a one-tailed hypothesis, the result really should not be considered as a demonstrable effect.

From the data we have, any difference between the therapies is unlikely to be meaningful.

Bayesian equivalent

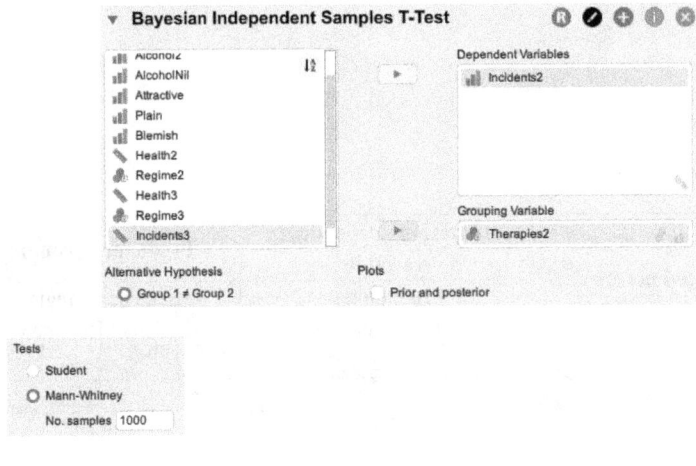

Bayesian Mann-Whitney U Test

	BF$_{10}$	W	Rhat
Incidents2	0.745	68.500	1.000

The Bayes factor is below 1, noise; if we had adopted a one-tailed hypothesis – Group 1 > Group 2 – the result would creep above 1, into the Weak credibility banding. *

Between-subjects one-way ANOVA: a parametric test for more than two conditions, different subjects

Let us extend our data from the Independent Samples *t* test example. The study previously examined different health scores for people under probation orders and in medium security prisons. We now think that perhaps we were not defining categories finely enough, so we examine a further sub-sample, confinement in an open prison.

---

*You may see some differences in the results each time the Bayesian Mann-Whitney is run. This is because of its underlying algorithm, which you will notice also takes some time to process. Greater stability in Bayes factor results may be achieved by raising the iterations (No. samples) from 1000 to 10,000 (Goss-Sampson, 2020).

Tests for different subjects

New variables Health3 and Regime3 have an additional 16 cases (27 to 42).

Sentenced person	Criterion / dependent variable	Predictor / grouping
	Score on test Name: **Health3**	Sentencing regime Name: **Regime3**
25	62	medium
26	71	medium
27	70	open
28	70	open
29	73	open
30	80	open
31	81	open
32	75	open
33	75	open
34	73	open
35	81	open
36	76	open
37	75	open
38	75	open
39	73	open
40	71	open
41	72	open
42	67	open

Press the ANOVA tab and select Classical / ANOVA from the menu.

Tests for different subjects

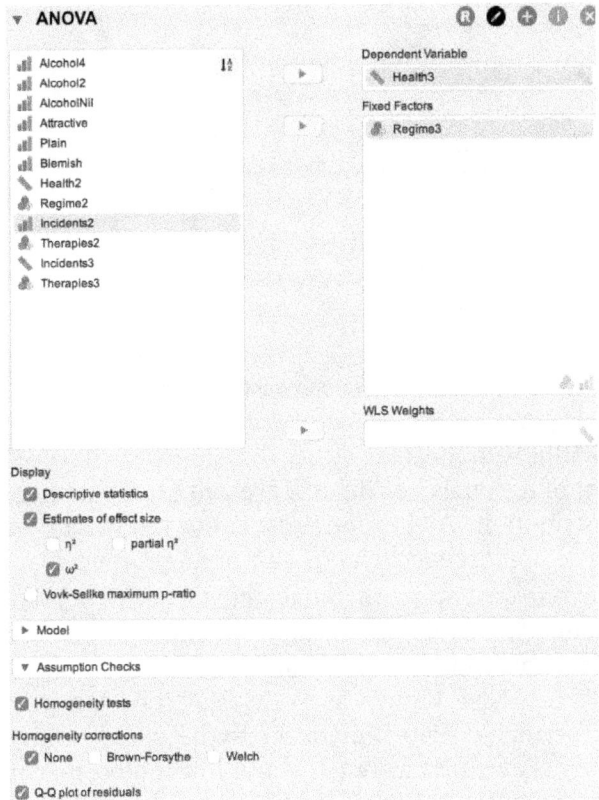

Transfer Health3 to the Dependent Variable box and Regime3 to the Fixed Factors box.

Descriptives

Descriptives - Health3

Regime3	Mean	SD	N
medium	63.333	6.286	12
open	74.188	3.987	16
probation	79.429	4.274	14

Assumption Checks

Test for Equality of Variances (Levene's)

F	df1	df2	p
2.575	2.000	39.000	0.089

Clear differences between the means of the conditions are suggestive of an effect. Levene's test is not significant.

Chapter 6 – Tests of differences

Tests for different subjects

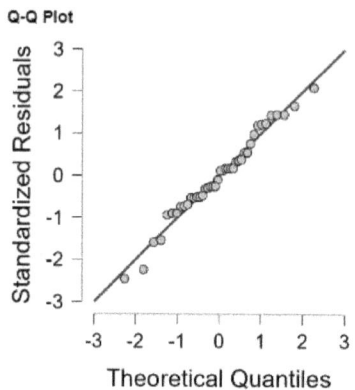

The points adhere quite closely to the slope of the Q-Q (quantile-quantile) plot of residuals (residuals represent error variance), indicating a normal distribution. As Levene's test is not significant, homogeneity (also known as homoscedasticity) is also not a problem.

If the Levene test were to be significant, with a p value smaller than .05, then you could go to the Assumption Checks panel and select a correction, which will adjust the ANOVA results. The Welch homogeneity test (also known as Welch's ANOVA), does not assume equality of variances. The Browne-Forsythe correction may be used in the case of skewed data, but it should be noted that this is a less powerful instrument than Welch, that is, less likely to detect an effect. If the data is clearly unsuitable, for example non-continuous, or without a normal distribution, then the non-parametric Kruskal-Wallis test should be used (introduced in the next section).

ANOVA - Health3

Cases	Sum of Squares	df	Mean Square	F	p	ω^2
Regime3	1721.301	2	860.650	36.863	< .001	0.631
Residuals	910.533	39	23.347			

Anyway, as we can continue, we see that the overall differences are significant. We have a large F ratio of 36.9 and p < .001. The larger the value of F, the easier it is to reject the null hypothesis.

You will also notice in the Display just below the variables boxes, the effect size measures eta squared (η^2), partial eta squared (η_p^2) and, my preference here, omega squared (ω^2). The results for eta squared and partial eta squared are the same in one-way ANOVA but differ with multiple ANOVA. Here, if you look, they give the figure .654. Although partial eta squared is a favorite in many textbooks, there is now evidence which indicates that it over-estimates the effect size in smaller samples, the bias being quite robust even in samples of 100 cases (Okada 2013). So my guess is that omega squared is more accurate here.

Omega squared shows us an effect size of .631, 63% of the explained variance, so we have a very large effect size. A rule of thumb for ANOVA effect size is that 0.01 is small, 0.06 is medium and 0.14 is large.

You may also choose Descriptive statistics. As well as showing the means and standard deviations, it includes the number of cases in each group, which in complex projects will tell you if you have set up your data correctly.

To see if these differences may be considered significant, Open Post Hoc tests. By transferring the grouping factor to the right, you can see the p value readings for pairings of levels.

We generally don't just run a series of t tests, because of the possibility of fluke results creating false positives (Type 1 error), so an adjustment is often made in the case of multiple comparisons of pairs. The default test in JASP is the Tukey test (a discussion of the choice of tests can be found in Chapter 11).

Open Descriptives Plots, transferring the fixed factor (here Regime3) to the right, onto 'Horizontal Axis'.

Tests for different subjects

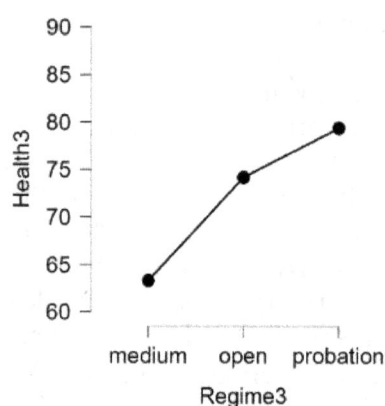

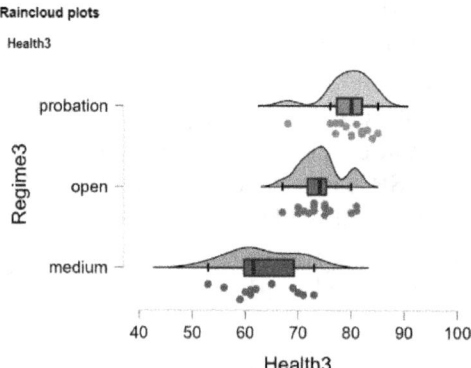

The Raincloud Plots section works in a similar way and comes out with a rather cool colour display (sorry, not here). Here, I have chosen the horizontal display option.

Tests for different subjects

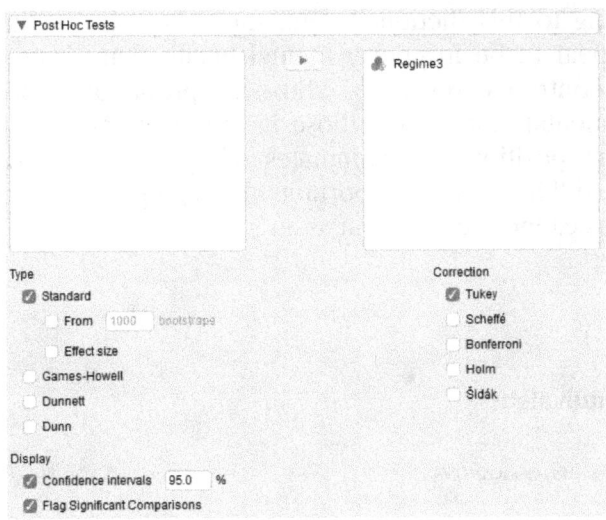

Post Hoc Tests

Standard

Post Hoc Comparisons - Regime3

		Mean Difference	95% CI for Mean Difference		SE	t	p_{tukey}
			Lower	Upper			
medium	open	−10.854	−15.350	−6.359	1.845	−5.882	< .001***
	probation	−16.095	−20.726	−11.464	1.901	−8.467	< .001***
open	probation	−5.241	−9.549	−0.933	1.768	−2.964	0.014*

* p < .05, ** p < .01, *** p < .001
Note. P-value and confidence intervals adjusted for comparing a family of 3 estimates (confidence intervals corrected using the tukey method).

From the post-hoc test statistics and from the charts, you can see the greatest effect when contrasting medium security prisoners and those on probation, and the smallest is between open prison inmates and probation. Note that the 'Standard' tests are in the vertical list, from Tukey down to Šidák. These and the more specialized tests on the left will be discussed in the second half of Chapter 11. I would merely note here that the Tukey test is the default test, and that as previously noted, the Bonferroni is widely used in the literature, but these days its correctional effects are considered rather harsh.

Tests for different subjects

According to this (fictional) data, those held in medium security prisons appear to be in poorer mental health than those in the less rigorously controlled régimes. Those on probation appear to have far better mental health than those in either prison condition. The intermediary position of the inmates of open prisons does suggest that regime differences are important, although pre-sentencing mental health status cannot be ruled out as an influential factor.

Bayesian equivalent

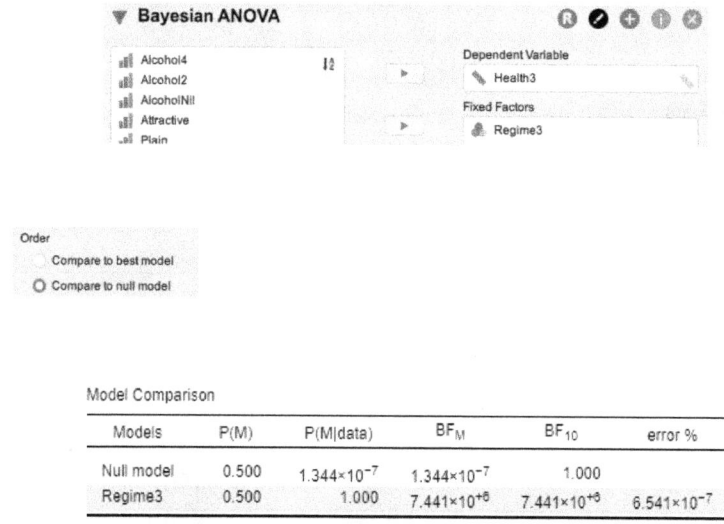

A huge BF10 is shown. It may, however, be best to see a plot to ensure that we have set things up properly.

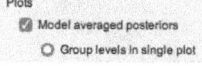

Tests for different subjects

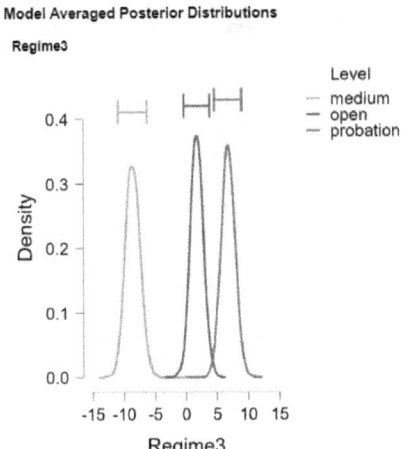

The posteriors are in distinctly different positions. On the other hand, there is some confluence between the central and right-hand posteriors, 'open' and 'probation' (these are in different colours in JASP).

Post Hoc Comparisons - Regime3			Prior Odds	Posterior Odds	$BF_{10, U}$	error %
medium	open		0.587	1178.875	2006.933	4.103×10^{-8}
	probation		0.587	89196.872	151850.037	1.453×10^{-8}
open	probation		0.587	11.902	20.262	3.269×10^{-6}

Note. The posterior odds have been Bayes corrected for multiple testing by fixing to 0.5 the prior probability that the null hypothesis holds across all comparisons (Westfall, Johnson, & Utts, 1997). Individual comparisons are based on the default t-test with a Cauchy (0, r = 1/sqrt(2)) prior. The "U" in the Bayes factor denotes that it is uncorrected.

There is some similarity to the classical version. While the first two pairings are extremely strong in terms of their credibility, the open–probation pairing has a Bayes factor of just over 20, edging into the Strong credibility banding.

Tests for different subjects

Kruskal-Wallis: a non-parametric test for more than two conditions, different subjects

If the data fails to meet the assumptions for ANOVA, continuous data and normality, use the Kruskal-Wallis test.

In addition to the two conditions examined by the Mann-Whitney test, two types of therapy, a fresh sample will come from another part of the country, where a third therapy (the excitingly-named C) is currently in use with patients with dementia. If you want to type in your own figures rather than use Differences.csv, the data for C are, 8 4 3 12 4   4 9 8 32 6

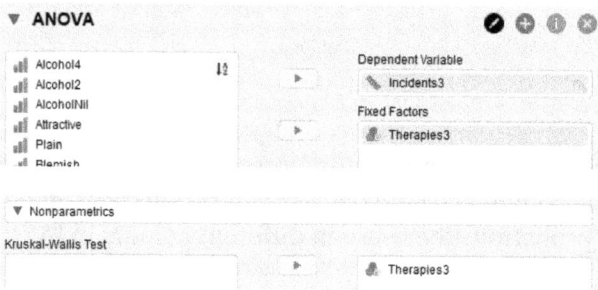

Press the ANOVA tab and select Classical / ANOVA from the menu. Incidents3 is the Dependent Variable; Therapies3 the fixed factor. The Nonparametics section runs the Kruskal-Wallis test.

Kruskal-Wallis Test

Factor	Statistic	df	p
Therapies3	1.790	2	0.409

The p value is about 0.41; clearly the null hypothesis cannot be rejected.

As I do not wish to dredge for data, we will not follow up with multiple comparisons. If we did wish to use a post-hoc test, Dunn's test, whick is non-parametric, would be appropriate. Hsu (1996) notes that it is possible for pairings to be significant even when the overall test is not. However, I would like to adopt a considered approach to post hoc tests rather than their mere availability; see the discussion on multiple comparisons in the second half of Chapter 11.

Let us turn to a data set where we know that there are significant results.

The study used for the ANOVA examined the results of mental health tests as taken in different types of criminal sentencing regimes. We need to set up the test as before, but this time use Health3 as the dependent variable and Regime3 as the grouping variable.

The result shows a p value of less than .001; we can also select the Dunn option. This shows highly significant results for all three pairings, although the medium security-probation pairing seems to enjoy even greater support for rejecting the null hypothesis.

*

Thinking point

It is a commonplace truism that correlations (Chapter 7) do not prove cause and effect. Experimental and quasi-experimental results are still subject to interpretation.

If we replicated our study of alcohol consumption and memory and found that lack of alcohol led to distinctly better recall in our experiments, would that mean that this was conclusive under other conditions (like not being woken in the middle of the night)? What about smaller measures of alcohol? Would there be differentials based on the importance or interest of the material being memorized?

Validity, results meaning what we think they mean, is missing. To improve our insights, triangulation is wise, carrying out different types of investigation in order to view the phenomenon from a different perspective. Fresh insights can sometimes lead to a complete rethink.

---

*At the time of writing, there is no Bayesian equivalent of the Kruskal-Wallis test.

Tests for different subjects

In the study involving mental health under different sentences, we might look at conditions within the different regimes, perhaps to see if different stress factors are at work. We could examine such relationships as that between contact with family members and depression; see Chapter 7 for correlations and regression. Another type of research may involve interviews or focus groups; see Chapter 8 for how qualitative results are sometimes quantified.

This table of tests of difference is not exhaustive, but provides general guidance.

N.b. *Non–parametric tests can be used with 'parametric' data, but in the author's view the reverse should not be happening.*

Design	Test	Conditions	Data
Same or paired subjects	Paired Samples T-Test	2	Parametric
	Wilcoxon signed rank	2	Non–parametric
	Repeated Measures ANOVA	3 or more	Parametric
	Friedman	3 or more	Non-parametric
Different Subjects	Independent Samples T-Test - Student's t	2	Parametric
	Welch's t	2	Parametric (good if equality of variances problem)
	Mann-Whitney U	2	Non–parametric
	ANOVA	3 or more	Parametric
	Kruskal–Wallis	3 or more	Non-parametric

All of these tests have a Bayesian equivalent test in JASP, except for Friedman, Kruskal-Wallis, and Welch.

Chapter 7 – Tests of relationships

Correlations

In this chapter, we are interested in the relationships between variables. There is a wealth of possible investigations. Issues coming to mind include possible relationships between incomes and crime, class and education, and the possible link between abortions and reduced levels of violent crime.

Statistically, the relationship is called **correlation**. The most used statistic is the **correlation coefficient**. It can run from 0 to 1, 0 being perfectly random and 1 representing a perfect positive relationship. It can also run from 0 to -1, -1 being a perfect negative relationship.

If there is a correlation between the introduction of abortions in the United States and violent crime, it is thought likely that this would be a negative one. The underlying theory is that abortions mean less unwanted children coming into the world, meaning less angry young adults, hence less violent crimes. You will find dissenting opinions on this subject, especially the vital point that correlations do not imply causation, but my point here is that statistical testing should have a rationale. If testing for correlations becomes a button-pressing exercise, just running tests on all the available data, misleading correlations become increasingly likely.

Correlations

As it is an important concept, let us dwell briefly on the non-causative nature of correlations. One variable may influence another, or it may not. The direction of causation may work in the opposite direction to what is expected. Or the relationship could be a horrible coincidence, with no causality whatsoever. But quite often, the explanation is elsewhere: there could be a mediating variable, one which explains the others.

The following examples will give us a practical start, using the file **Correlations.csv**:

PerfPosA	PerfPosB	PerfNegA	PerfNegB
1	1	1	5
2	2	2	4
3	3	3	3
4	4	4	2
5	5	5	1

The first two variables in the Correlations.csv file have a perfect positive relationship with each other. The second pair of variables forms a perfect negative relationship. These examples do not occur in real life, but are here for demonstration purposes.

Press the Regression tab, select Classical / Correlation from the drop-down menu and transfer the relevant variables to the right-hand box. Keep the default options, including Pearson's r (to be discussed), Correlated, and Report significance but also choose Scatter plots from the Plots section in order to see the chart.

Correlations

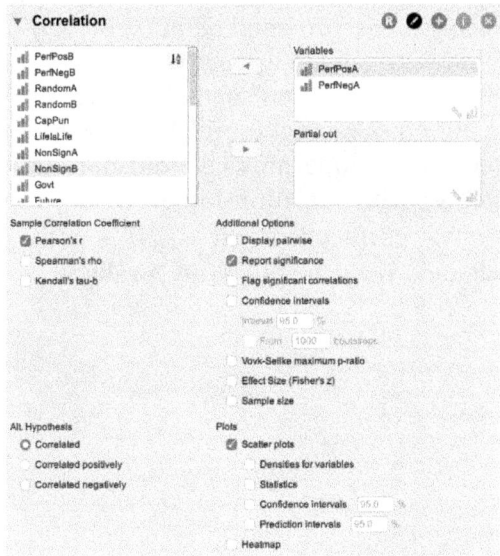

Here, we have two identical sub-samples, both 1, 2, 3, 4, 5.

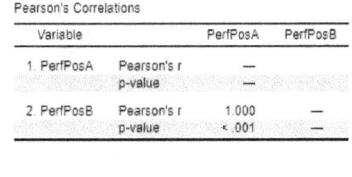

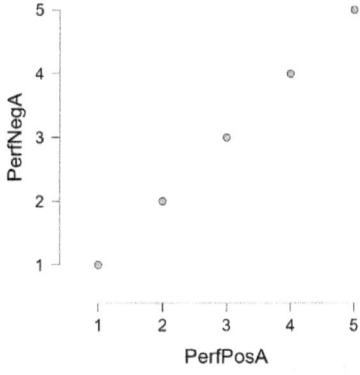

Correlations

As well as producing a Pearson coefficient (*r*, or *rho*) of 1, the perfect relationship, with a very small *p* value, you can also see a scatter plot. Note that to get a rather more presentable chart, I have selected the Display pairwise option. A perfect correlation, *r* = 1, does not generally occur in research, but a slope moving from bottom left to top right generally indicates a positive relationship between two variables.

Now place the two perfect negative variables in the right-hand box. Pearson's *r* should be - 1, again with a tiny *p* value.

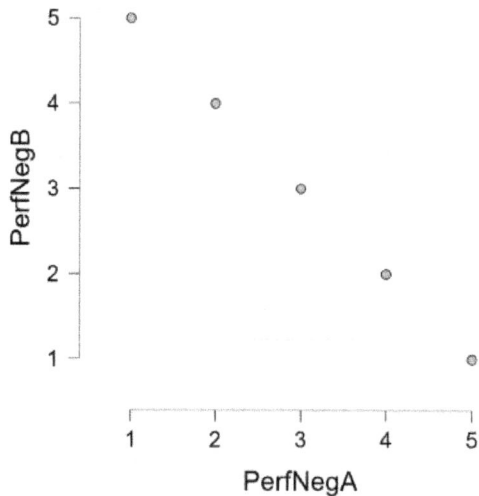

A slope rising from bottom right to top left indicates a negative relationship between two variables.

Now use the two randomly generated variables.

Correlations

RandomA	RandomB
80	83
10	70
84	79
42	98
13	62
76	12
28	29
97	87
12	62
98	44

Firstly, note that the points representing pairings of data are scattered about the chart in a globular cluster. The roughly horizontal line through the data is also characteristic of the absence of a correlation, as is the remoteness of much of the data from the line.

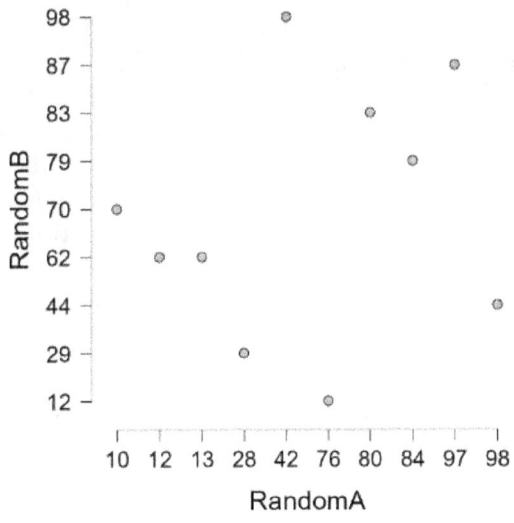

Correlations

Pearson's *r* is 0.035, very close to zero. The *p* value, 0.923 is very high, close to 1. In general, a low *r* denotes the absence of a relationship. Similarly, the high *p* value indicates lack of support for rejection of the null hypothesis; the null hypothesis is that there is no relationship between the variables.

Correlations and effect sizes

Let us consider a few key terms. The **correlation coefficient**, *r* (short for rho) in the case of the Pearson and Spearman tests, *tau* for Kendall's rank correlation, is a measure of the strength of the relationship in a particular direction (positive or negative).

$-1 \Leftarrow 0 \Rightarrow 1$

A coefficient of 1 means a perfect relationship, -1 means a perfect inverse (or 'negative') relationship, with 0 as the ideal random relationship. So for a positive relationship between variables, the nearer to 1, the stronger it is; a coefficient of .924, for example, is clearly very strong in the positive direction. Similarly, a coefficient of -.924 would also be strong, but in the negative sense.

Before deciding that a positive correlation means a positive relationship between a pair or variables, or that a negative represents a conceptual inverse, it is well worth studying your data (as is recommended generally in this book). A common situation arises in questionnaire development, where scales run in different directions to avoid response order effects (for example, people always picking a high number regardless of the content of the question). The apparent direction of a relationship may be an artefact of the scales you are using.

The question of the size of a correlation is a vexatious one. I cite positive sizes in the next two paragraphs, but you should consider negatives as the exact inverse.

For smallish data sets, one could say that pairings of variables with a correlation coefficient of between .9 and 1 should be considered as very highly correlated; those of a magnitude of between .7 and .9 are highly correlated; those between .5 and .7 are moderately correlated; .3 to .5

are low correlations. Below .3, the relationship is weak or non-existent (Calkins, 2005).

For larger data sets, it is more reasonable to consider anywhere between .5 and 1 as large, .3 to .5 as medium and .1 to .3 as small. It is also worthwhile looking at reports of similar studies and, when dealing with 'big data', effect sizes are likely to be more helpful than correlations or p values, as they deal with the variance.

The p value is our guide to how likely it is that the effect is 'real', that is, significant. The null hypothesis for correlations is that there is no reason to believe that a relationship exists; the lower the p value, the more likely it is that we can reject the null hypothesis. Sometimes, as in our perfectly negative and positive examples, you will see significance shown as something like $p < .001$. That doesn't mean that JASP has set a critical value (see Chapter 4); it means that the value of p is just too small to reproduce accurately.

The p value is particularly suspect when dealing with 'big data'. Lots of relationships will appear to be significant, often truthfully with that amount of evidence. The question then becomes just how meaningful are the relationships, and I would definitely extend this question to smaller data sets as well.

Coefficients and p values are sometimes mistaken for the strength of the effect. This is in fact represented by the **effect size**. A measure of the variance, this term is rarely seen in traditional books on statistical testing but is now increasingly viewed as important, particularly when dealing with big data. Essentially, one may be happy that a result is statistically significant, that an effect exists, but the effect size comments on the extent of its influence, its magnitude.

There is a range of methods for calculating effect size. By default, for ease of use, I suggest the squaring of the correlation statistic.* So if you have r from the Pearson test, you would calculate r squared (or r^2), multiplying r by itself. So if r is .73, the r squared is $.73 \times .73 = .53$, just over half of the variance. Note that if we square a minus, in this case -.73, we still end up with .53; the strength of the effect is independent

---

*Kendall's *tau* should not be squared, however.

Correlations

of direction. (Do test your calculator with the examples just given, as some calculators can't square negatives. If you choose to continue with a problem calculator and have to square negatives, you can just square positive values for the same result; .73 × .73 is the same as -.73 × -.73.)

There are no agreed reporting categories for effect sizes. Cohen (1977) recommends:

Large: .8   Moderate: .5   Small: .2

If in doubt, examine the results of similar studies.

The Pearson test: a parametric correlational test

CapPun	LifeIsLife
62	65
48	52
44	39
37	47
62	66
54	54
68	73
55	58
68	72
60	64

Groups of respondents in 10 different areas are asked about their attitudes to penal policy. Each group is represented by a percentage in favor of different statements. Here, we are interested in the relationship between those in favor of capital punishment and those agreeing with the statement 'Life means life' (that is, that a 'life sentence' means actual physical imprisonment for the whole of a prisoner's lifetime).

As previously, press the Regression tab, select Classic/Correlation from the drop-down menu. Transfer the CapPun and LifeIsLife variables to the right-hand box.

The Pearson test is a parametric test and we need to check that the data meets the assumptions for such a test. First, go to the Plots section and choose Scatter plots. This provides the correlation chart, allowing you to check for linearity (of which more later) as well as to spot any

Correlations

outliers. You can also select Densities for variables, which shows if you have something akin to normal distribution (I have used small samples, which make for less than beautiful viewing). Another thing you can do is to press the Descriptives tab and, in the Statistics panel, Distribution section, use the Shapiro-Wilk normality test (which is not significant in this example, indicating normality). *

Kim (2013) reports that formal tests such as Shapiro-Wilk are usable for samples smaller than 300, but may become unreliable for larger data sets. Another option, again using the Descriptives tab, Statistics panel, is to examine the Skewness and Kurtosis distribution statistics; the nearer to zero for each, the better. Between +1 and -1 are perfectly reasonable figures. Although there are no official limits, many statisticians prefer the figures to be within the bounds of +2 and -2. West *et al* (1996) suggest a maximum of 2 for skewness and 7 for kurtosis. Kim (2013) recommends a limit of 1.96 for kurtosis and skewness for samples of less than 50; over 3.29 for samples of from 50 to 300; and for more than 300, go back to the correlation matrix plot and follow the advice of West *et al*.

With unsuitable data, we would need to use the Spearman's rho or Kendall's tau-b tests of correlation, as demonstrated in the next section.

As we have normally distributed data, we can use the default Pearson test. Here we can see the scatter plot. A clear positive slope is to be seen: most of the data gathers quite closely to the slope.

---

*I don't use the Pairwise version of Shapiro-Wilk in Assumption Checks section of the Correlation dialog. The Descriptives version is in line with other software such as SPSS.

Correlations

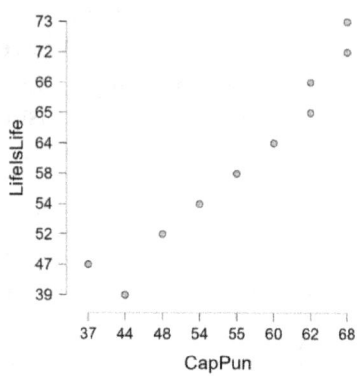

Pearson's *r* is .939, with an effect size of .882 (.939 squared *), approaching 89% of the variance. This means that only about 11% of the variance from the mean is likely to be due to chance or additional factors.

Use the Descriptives tab to check the means, 55.8 and 59.0, not very much different. Also, if you go to the Dispersion section of Descriptives and select Range, which tells you the difference between the highest and lowest figures in a data set, you see similar ranges. This suggests, but does not prove, a correlation. Again, these are useful statistics for reporting (as is standard deviation).

Let's see a result where we cannot reject the null hypothesis:

---

*I do not use the Fisher's *z* effect size option.

Correlations

NonSignA	NonSignB
52	60
53	34
47	38
40	52
48	54
45	55
52	36
47	48
51	44
38	56

Pearson's r is a minus number, -0.488, and the p value at 0.153 is quite high. The scatter plot shows much of the data deviating from the slope. This is a non-significant result.

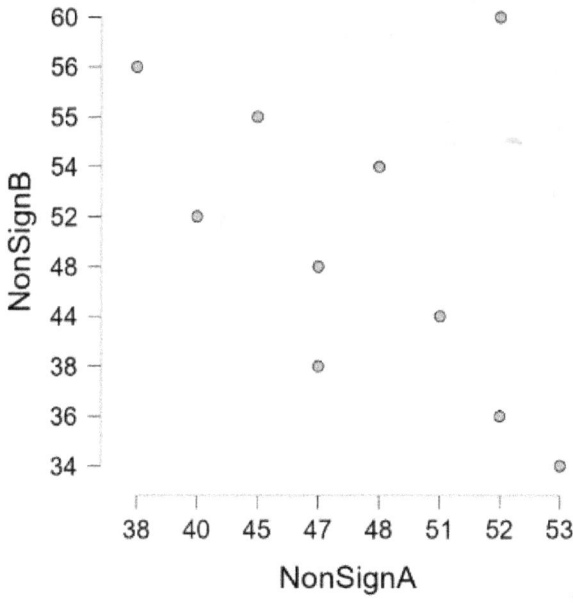

Correlations

Bayesian equivalent

Statistic		Quantification of evidence
Bayes Factor (BF10)	BF reciprocal (BF01)	
< 1	> 1	Noise
1 – 3	1 – 0.33	Weak
3 – 10	0.33 – 0.1	Moderate
10 – 20	0.1 – .05	Positive
20 – 150	.05 – .0067	Strong
> 150	< .0067	Very strong

This is a reminder of a suggested guide to Bayesian hypothesis reporting. Open Regression / Bayesian / Correlation.

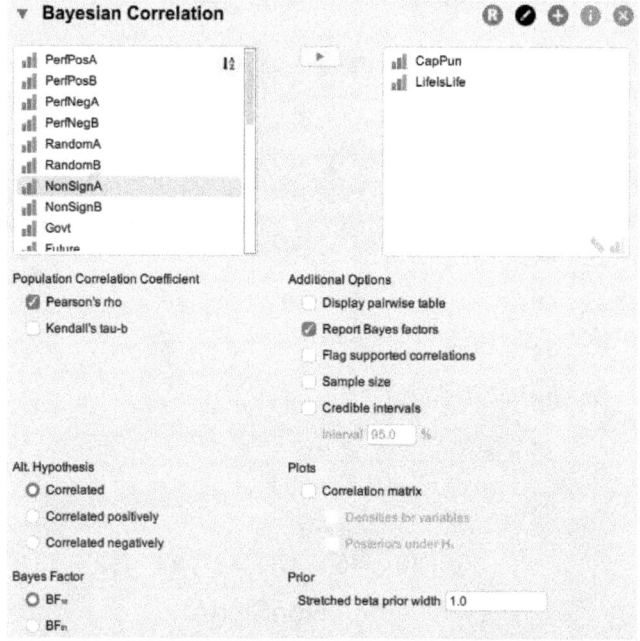

Correlations

Bayesian Pearson Correlations			
Variable		CapPun	LifeIsLife
1. CapPun	Pearson's r	—	
	BF_{10}	—	
2. LifeIsLife	Pearson's r	0.939	—
	BF_{10}	369.490	—

Pearson's *r* is reproduced, accompanied by a Bayes factor, showing very strong support (greater than 150) for the alternative hypothesis.

Here we show the result using variables NonSignA and NonSignB:

Bayesian Pearson Correlations			
Variable		NonSignA	NonSignB
1. NonSignA	Pearson's r	—	
	BF_{10}	—	
2. NonSignB	Pearson's r	−0.488	—
	BF_{10}	0.960	—

The Bayes factor falls short of 1, so is considered to be 'noise'.

The Spearman and Kendall's *tau-b* tests: non-parametric correlational tests

Spearman is used in much of the traditional literature. These days Kendall's *tau-b* is becoming more popular. There is generally little difference in their interpretation of results, although it has been said that Kendall is more accurate for smaller samples (particularly less than 12) and ones with more tied ranks, while Spearman is better for nominal categories (for example, city 1, city 2 ... city x) with no hierarchical ordering. Do note that the statistical community (as usual) is divided over their respective qualities, but Kendall is now generally preferred.
* Spearman's pervasiveness in much of the literature is because in the days before the personal computer, it was much quicker to calculate. It is also said by the cynical that the Spearman test is more popular because Spearman's *rho* is usually larger than Kendall's *tau*.

---

*You can't square Kendall's *tau* for effect size however.

Correlations

If your data set fails the Shapiro-Wilk test, it is likely to have a non-normal distribution and is thus unsuitable for the Pearson test. In the Correlation Coefficients section, choose Spearman or Kendall's *tau-b* instead of Pearson.

In a survey, people express their level of confidence in the government, with a rating of 1 for very low to 5 for very high. Another scale measures confidence in the future, ranging from 1 for very much lacking in confidence in the future to 5 for very confident in the future.

The scale 'Confidence in the Government' (running at low ebb, it would seem) is:

1, 5, 4, 2, 2, 3, 1, 1, 3, 2

The scale 'Confidence in the Future' (uncertainty rules) is:

5, 4, 5, 3, 1, 2, 4, 3, 2, 2

The variables in Correlations.csv have been named Govt and Future.

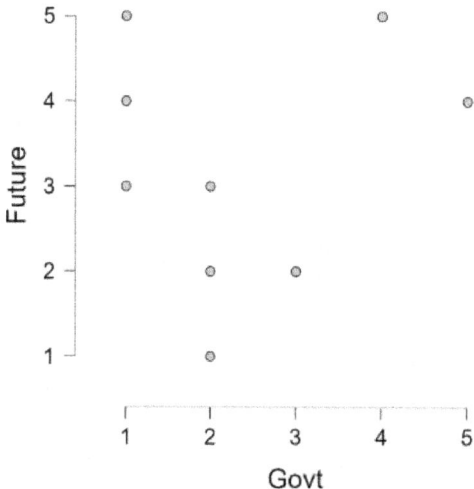

The Spearman coefficient is about –.06, obviously small, with a *p* value of 0.875, very big; any effect is clearly a matter of chance. Kendall's *tau-b* readings are quite similar.

Correlations

Now compare the 'Confidence in the Future' ratings with this Salaries variable (in thousands for each person):
28, 30, 25, 27, 18, 20, 15, 24, 18, 22.

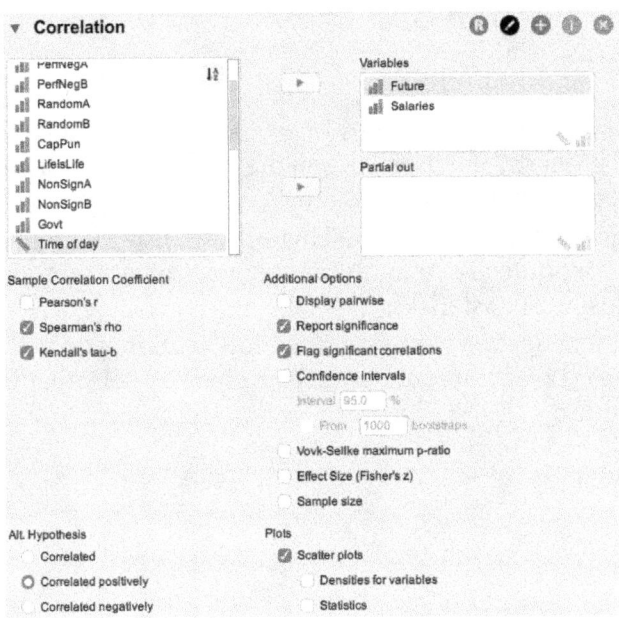

Let us assume previous evidence to suggest that the well-heeled are generally more confident. We may have considered it likely that salary would be positively associated with confidence in the future. So we select the one-tailed hypothesis, selecting Correlated positively from the Alt. Hypothesis section and, for clarity, Flag significant correlations.

Correlations

Correlation Table

Variable		Future	Salaries
1. Future	Spearman's rho	—	
	p-value	—	
	Kendall's Tau B	—	
	p-value	—	
2. Salaries	Spearman's rho	0.572*	—
	p-value	0.042	—
	Kendall's Tau B	0.483*	—
	p-value	0.032	—

Note. All tests one-tailed, for positive correlation.
* $p < .05$, ** $p < .01$, *** $p < .001$, one-tailed

As we consider the less rigorous one-tailed level to be acceptable on theoretical grounds, and have chosen the criterion level of $p < .05$, then we can reject the null hypothesis.

If we had chosen the two-tailed hypothesis, using merely Correlated in the Alt. Hypothesis section, then we would have seen the p values doubling, losing our asterisks.

If you select Scatter plots from the Plots section, you will see a mild slope. (Here, I go for a more attractive chart by selecting the Display pairwise option.)

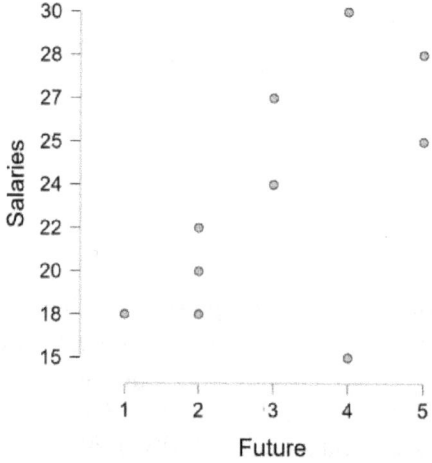

Correlations

Bayesian equivalent

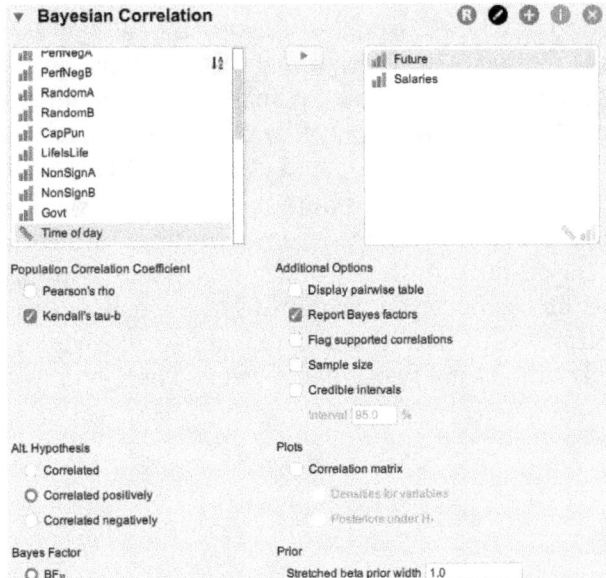

JASP produces a Bayesian equivalent to Kendall's *tau-b* test. Open Regression / Bayesian / Correlation.

Bayesian Kendall's Tau Correlations

Variable		Future	Salaries
1. Future	Kendall's tau	—	
	BF₊₀	—	
2. Salaries	Kendall's tau	0.483	—
	BF₊₀	4.014	—

Note. For all tests, the alternative hypothesis specifies that the correlation is positive.

The Bayes factor for the one-tailed hypothesis falls within the Moderate credibility banding (3 – 10). If we had opted for the two-tailed version, it would have been within the Weak banding.

Correlations

A cautionary note

It is a very good practice to examine a scatter plot as well as using a test. The reason is that tests of correlational relationships, non–parametric as well as parametric, are linear. A relationship essentially runs in a straight line. Assuming significance or non-significance without using a scatter plot could be a big mistake. Let us look at two examples.

One error was made by the author (reader gasps). I examined a friend's blood pressure readings against the time of day (thanks, you know who, for permission to reproduce this evidence of my impulsive nature). We had expected a relationship between the two variables and were surprised to find a non-significant result, with a coefficient of -0.16 and with .3 for a two-tailed p value. Then, it dawned on me.

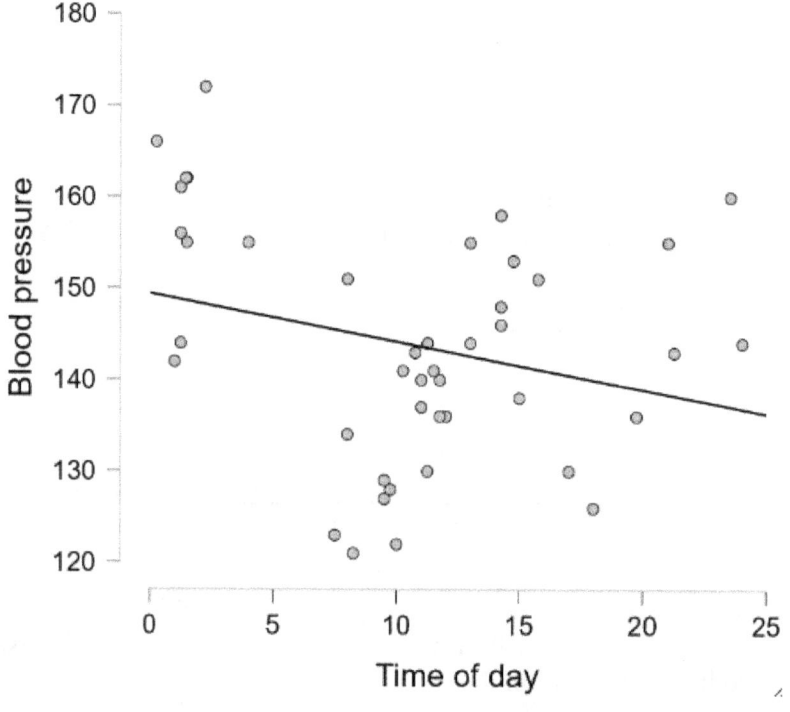

This expanded image of the chart shows something like a W shape in the data far away from the slope. My friend tended to have higher blood pressure readings in the early afternoon and at night. The effect we show here is not linear. The proliferation of data well away from the shaded area is a tell-tale sign. A test assuming linearity was of course meaningless, as was the projected slope.

Now, for students under stress and sports fans, I present a classic non-linear relationship, between performance and arousal:

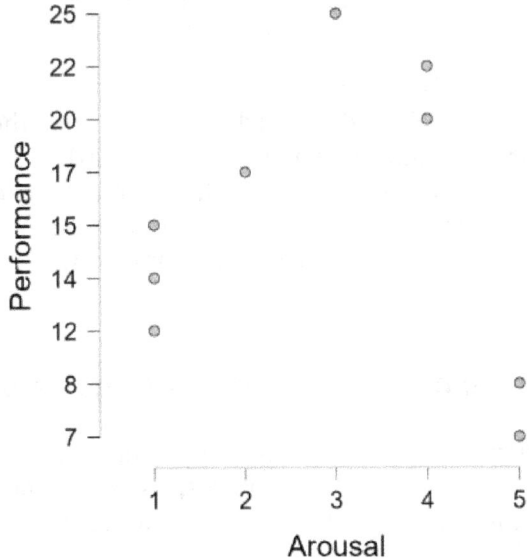

The loose data appears as an inverted U shape. This is a crude example of the Yerkes-Dodson Law (Yerkes and Dodson, 1908). Some degree of stimulation appears necessary for performance, but high stress and low levels of arousal may damage performance. The non-linear effect emerges from the chart, but correlational tests, parametric or non-parametric, would be inappropriate. These tests are not appropriate for 'curvilinear' relationships. You should always examine your data visually before considering test results.

Correlations

If we return to our main parametric example with a significant result, the relationship between support for capital punishment and for whole-life prison terms, a correlation coefficient of .939 when squared gave .882, a large effect size. Now consider the relationship between the variables indicating Satisfaction with life and Income (fictional data).

Pearson's Correlations

			Pearson's r	p
Satisfaction	-	Income	0.725**	0.009

Note. All tests one-tailed, for positive correlation.
* $p < .05$, ** $p < .01$, *** $p < .001$, one-tailed

We have reason to reject the null hypothesis, that there is no relationship, with a p value that is smaller than .01, but squaring the coefficient gives us an effect size of .53, not much more than half of the variance. One way of examining a phenomenon more broadly is to look at correlations of more than a single pair of variables.

Multiple correlations – parametric - using Pearson's test

Thus far, we have focused on relationships between two variables. It is quite possible to use several variables, as we do here. Do note, however, that the more correlations you include, the more likely that some apparently significant relationships are in fact due to chance (Type 1 errors). So try to have a rationale for including each variable.

On this occasion, we are interested in a (fictional) study of social influences on children. We are concerned about violence between groups of young people in their early teens. Our particular interests are the connection between watching violent videos and actual violence, and the possibility that politicization is positively related to violence in the particular area under study. As well as measures of video watching, politicization and violence, we also include measures of educational attainment and economic status.

Correlations

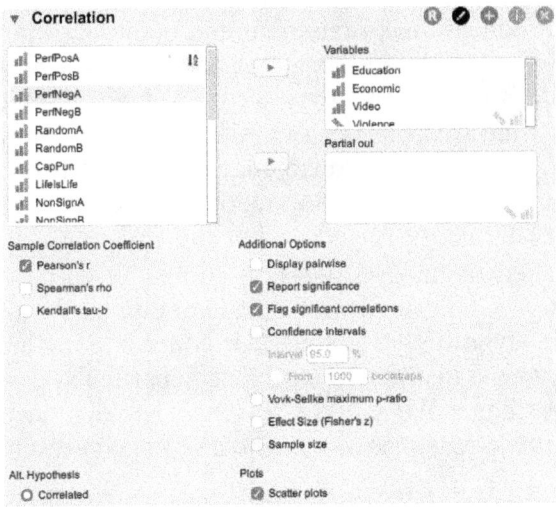

Press the Regression tab, select Classical / Correlation from the drop-down menu and transfer the following variables to the right-hand box: Education, Economic, Video, Violence, Politicisation.

If you choose the Scatter plots from the Plots section, you will see a grid of scatter plots for each pairing (larger grids can be created by opting for Display pairwise. For the correlation matrix itself, let us assume that you have started with the default two-tailed hypothesis, Correlated, which is without a predicted direction. The Flag significant correlations option makes it easier to detect significant results; asterisks appear on the table of results.

Pearson's Correlations

Variable		Education	Economic	Video	Violence	Politicisation
1. Education	Pearson's r	—				
	p-value	—				
2. Economic	Pearson's r	0.743**	—			
	p-value	0.009	—			
3. Video	Pearson's r	−0.218	−0.260	—		
	p-value	0.520	0.440	—		
4. Violence	Pearson's r	−0.405	−0.634*	0.492	—	
	p-value	0.217	0.036	0.124	—	
5. Politicisation	Pearson's r	0.355	−0.025	0.332	0.600	—
	p-value	0.284	0.941	0.318	0.051	—

* p < .05, ** p < .01, *** p < .001

Chapter 7 – Tests of relationships

Correlations

It is clear that educational attainment and economic status are highly correlated, with a moderate negative correlation for economic status and violence. In both cases, they are considered significant. This is not the case when we consider the relationships of interest, although the pairing of politicization and violence is moderately correlated, with a lowish correlation pertaining to videos and violence, and also a low negative correlation between education and violence.

Now we need to make a decision about the correct use of one-tailed hypotheses. This needs to be made before taking a peek at other options in the Hypothesis section. Our concern combines the theory (or rationale) with statistics: Do we expect a particular direction for the pairing? If yes, then a one-tailed test is reasonable; if no, then it is not. (One unacceptable reason is just to find a significant result.)

Pearson's Correlations

Variable		Education	Economic	Video	Violence	Politicisation
1. Education	Pearson's r	—				
	p-value	—				
2. Economic	Pearson's r	0.743**	—			
	p-value	0.004	—			
3. Video	Pearson's r	-0.218	-0.260	—		
	p-value	0.740	0.780	—		
4. Violence	Pearson's r	-0.405	-0.634	0.492	—	
	p-value	0.891	0.982	0.062	—	
5. Politicisation	Pearson's r	0.355	-0.025	0.332	0.600*	—
	p-value	0.142	0.530	0.159	0.026	—

Note. All tests one-tailed, for positive correlation.
* $p < .05$, ** $p < .01$, *** $p < .001$, one-tailed

Our rationale for the study was that violence and politicization are likely to be positively related (I mean positively in the statistical rather than the moral sense), so at least for this relationship, we consult the table using a unidirectional hypothesis, Correlated positively.

Do note that when you look at the tables through the lens of a particular hypothesis, it will apply to all of the pairings, which obviously does not make sense. So when reporting, it is best to report each pairing separately, referring to 'one-tailed' or 'two-tailed' in each case.

Violence and Politicisation are considered to be significantly related. Although the rejection of the null hypothesis in the previous matrix

Correlations

because of a p value of 0.051 is of course questionable. The critical value of $p < .05$ was of course Fisher's rule of thumb. You may want to look at this data now using Bayesian analysis, which gives a graded set of reported results rather than a 'significant or non-significant' result.

Bayesian equivalent

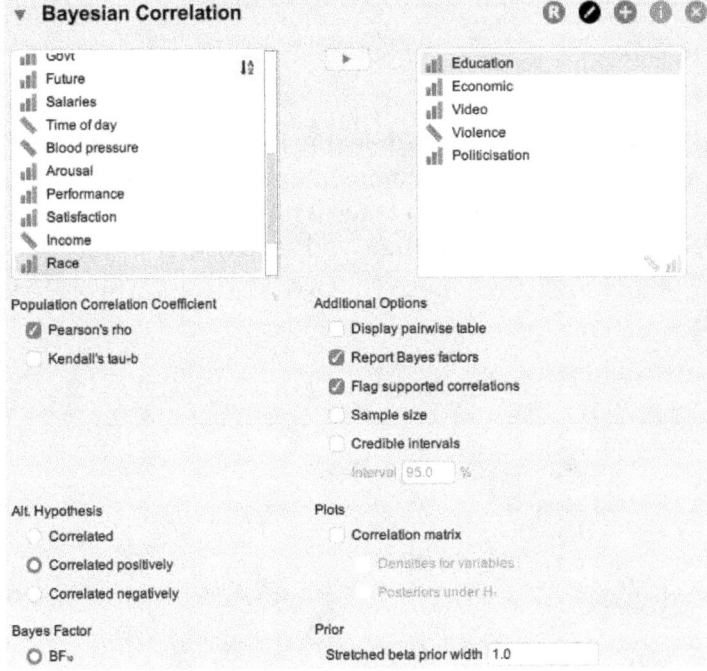

Here, we have opted for the one-tailed hypothesis, 'Correlated positively', and as we are examining a table of correlations, the option 'Flag supported correlations'.

Correlations

Bayesian Pearson Correlations

Variable		Education	Economic	Video	Violence	Politicisation
1. Education	Pearson's r	—				
	BF_{+0}	—				
2. Economic	Pearson's r	0.743*	—			
	BF_{+0}	15.152	—			
3. Video	Pearson's r	−0.218	−0.260	—		
	BF_{+0}	0.246	0.231	—		
4. Violence	Pearson's r	−0.405	−0.634	0.492	—	
	BF_{+0}	0.188	0.143	1.965	—	
5. Politicisation	Pearson's r	0.355	−0.025	0.332	0.600	—
	BF_{+0}	1.037	0.350	0.951	3.883	—

* $BF_{+0} > 10$, ** $BF_{+0} > 30$, *** $BF_{+0} > 100$
Note. For all tests, the alternative hypothesis specifies that the correlation is positive.

JASP only flags up correlations where the Bayes factor is over 10 (having at least a 'Positive' credibility banding). So, while Economic–Education is recognized, Politicisation–Violence is not. Leaving aside the flagging, the Bayes factor for politicization and violence is in the 'Moderate' banding.

Bayesian Pearson Correlations

Variable		Education	Economic	Video	Violence	Politicisation
1. Education	Pearson's r	—				
	BF_{10}	—				
2. Economic	Pearson's r	0.743	—			
	BF_{10}	7.640	—			
3. Video	Pearson's r	−0.218	−0.260	—		
	BF_{10}	0.445	0.483	—		
4. Violence	Pearson's r	−0.405	−0.634	0.492	—	
	BF_{10}	0.732	2.607	1.066	—	
5. Politicisation	Pearson's r	0.355	−0.025	0.332	0.600	—
	BF_{10}	0.619	0.370	0.579	2.016	—

Here, I have opted for the two-tailed, 'Correlated', hypothesis. The politicization and violence pairing are now in the 'weak' banding.

Multiple correlations – non-parametric - using Spearman/Kendall's *tau-b*

Press the Regression tab and select Classical / Correlation from the drop-down menu. Transfer the relevant variables to the right-hand box: Race, Aid, Hawkish, Immigration.

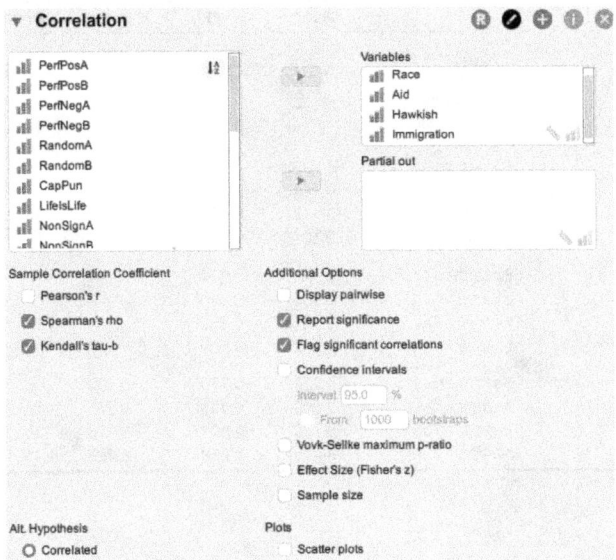

The interest here is in social views, using uncalibrated 5-point Likert scales. If you checked normality by going to Paired T-Tests with these variables, you would find that most do not fail the Shapiro-Wilk test (that is, the results are not significant). However, I am wary of treating scales as if they comprised continuous data. The basic rule is that if the concept can't really be halved, for example that half of a 4 rating really is not equivalent in meaning to a 2, then it should not be treated as truly continuous.

There are other reasons why you might not want a parametric test. Perhaps the variables are of differing types of data. Perhaps you do not think they are close to normal distribution because of the graph or

Correlations

skewness or kurtosis. Sometimes, as I said earlier, I think you just have to look at the type of data you are putting in and make a logical decision.

I have opted for the Flag significant correlations option, which will show us asterisks.

Correlation Table

Variable		Race	Aid	Hawkish	Immigration
1. Race	Spearman's rho	—			
	p-value	—			
	Kendall's Tau B	—			
	p-value	—			
2. Aid	Spearman's rho	-0.233	—		
	p-value	0.491	—		
	Kendall's Tau B	-0.172	—		
	p-value	0.507	—		
3. Hawkish	Spearman's rho	-0.341	0.208	—	
	p-value	0.304	0.540	—	
	Kendall's Tau B	-0.211	0.170	—	
	p-value	0.413	0.509	—	
4. Immigration	Spearman's rho	-0.228	-0.776**	-0.190	—
	p-value	0.500	0.005	0.575	—
	Kendall's Tau B	-0.200	-0.697**	-0.176	—
	p-value	0.445	0.008	0.499	—

* $p < .05$, ** $p < .01$, *** $p < .001$

The correlations table only picks up on one relationship, a negative one between immigration and aid. We can feel reasonably sure that we can reject the null hypothesis that the effect does not exist, but how important is the effect in terms of real life? If we square Spearman's *rho* (-.776 × -.776), we get an effect size of .602, 60% of the variance.

Bayesian equivalent

The Bayes factor for the immigration–aid pairing is flagged; at 19.113, this is within the 'Positive' credibility banding (and just short of 'Strong').

Thinking point

Various ideas emerge from this. If the rest of the variance is random 'noise', how can our research model be improved? Is there another influential variable which would help us to better understand the effect and create a better model? Is the new model relevant and should we invest time and resources to it? These questions become clearer when we consider multiple regression.

As a building block, however, we first need to consider regression as a concept.

Regression

In general terms, regression is a predictive tool used to create statistical models. A model is not intended to be an accurate reflection of the world, but an idealization. It creates a usable set of concepts to help us handle a problem or theory in a practical way. Multiple regression is a major tool in this process, but in order to understand multiple regression, we need to consider the concepts within simple linear regression. We will also be building upon our knowledge of correlations.

Simple linear regression (two conditions) – parametric

Regression appears on the face of it to be similar to correlations. We are still interested in the nature of relationships between data and we are still interested in a slope, although this is known in regression as *the line of best fit*. Both methods assume a normal distribution, although arguably this is more important for the dependent variable in regression. Both require linear relationships.

There are some differences. The mathematical method is different. Although I will not cite the equations, you do need to understand what regression does, and with what type of variable. We will examine it here by comparing it with the already familiar concept of correlations.

Regression

Correlations examine a mutual relationship between variables, with no mathematical differentiation between the variables. We only know that they are related to each other. Any choice of direction, with a one-tailed hypothesis, is a matter for the test user; any assumption of cause and effect is at the researcher's own risk.

Linear regression assumes that one variable affects another, assuming one direction (if you ever go to another book, you will find two formulae, one for how x influences y and another for y influencing x, but you don't have to go there). So the dependent variable, the one being affected, needs to be a continuous variable which can be acted upon. For example, we might want to consider salaries as a dependent variable, or the number of incidents of a particular nature, or examination scores. One relevant question of this type is whether or not the number of years in education affects income. Another is whether or not levels of policing have an effect on criminal incidents. Yet another would be the effect of hours of television-watching on examination scores.

One important feature of regression is that the relationship between the two variables can lead to predictions. When we have a line of best fit in simple regression, which involves only two variables, this can be extended to extrapolate beyond the line. Numerically, we can see how far one unit of the predictor (referred to as the covariate in JASP) produces an increase in the criterion variable (dependent variable). For example, the regression could indicate that for every additional year in education, subsequent incomes rise by a certain amount. For each hour of television watching, students' examination scores could be reduced on average by so many points (fictional data as usual). In other software and some of the literature, the predictor is sometimes known as x (as in the horizontal axis of a chart) and the dependent variable as y (the vertical axis).

As usual, some common sense is required. In the case of policing, it is quite likely that the number of incidents affects policing levels as well as vice-versa. So although regression is used with the aim of describing cause and effect, it still needs to be applied logically.

Let us look at a practical example. Use the **Regression.csv** file.

Regression

Firstly, it helps to check that there are relationships between the data. If you open up Regression / Classical / Correlation and examine the variables Education and Income with the correlation plot, you will see a clear, strong, linear relationship. If our data refers to the educational level of individuals and their subsequent earnings, then it is clearly worth investigating Education as an explanatory factor, influencing dependent variable Income.

However, if Income means parental income, more likely to be a causal variable, we may need to reverse our variables, making Education the dependent variable – if you experiment with the correlation matrix plots and reverse the order of the variables, you will find a subtle difference in the scatter plots. Whereas the correlation coefficients are not affected by the order in which the variables are placed, the plot is different, which is important when applied to regression. Regression assumes causation, unlike correlations.

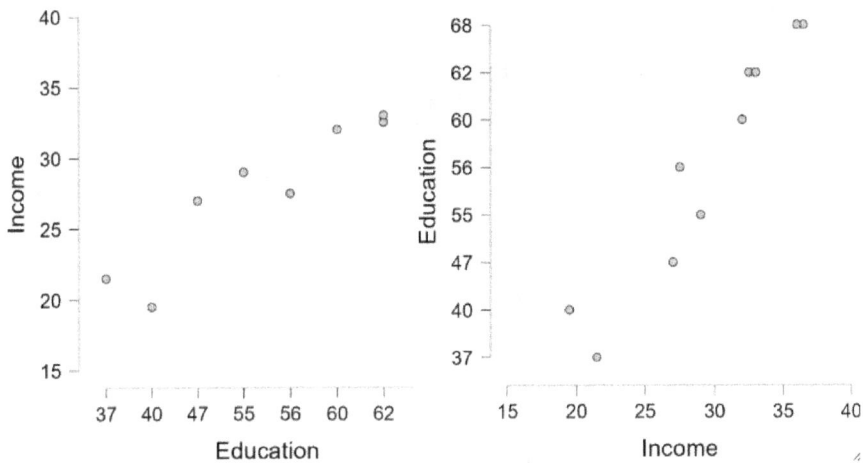

Press the Regression tab, selecting Classical / Linear Regression from the drop-down menu. Income fits into the Dependent variable box, with Education transferred to Covariates. So, we expect education to influence future earnings.

Regression

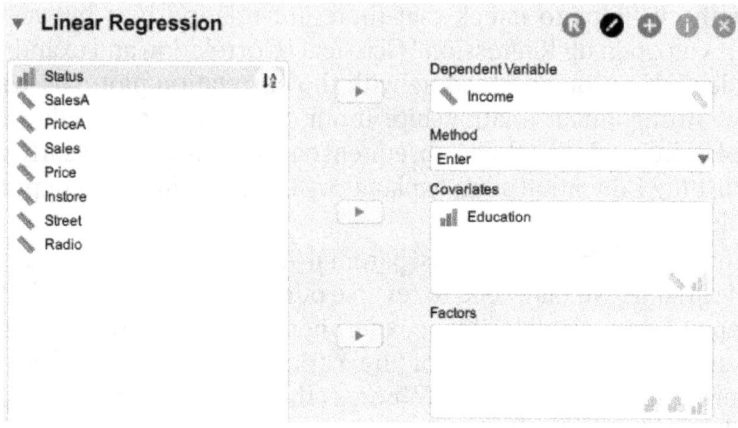

N.b. Do not alter the Method drop-down box; it should remain at its default, 'Enter'. (The other options are discussed briefly in Chapter 12, under the title 'Sequential regression'.)

Education	Income
62	32.5
56	27.5
40	19.5
37	21.5
62	33
47	27
68	36.5
55	29
68	36
60	32

Here are the data for this example. Do note that Income is in thousands (e.g. 32.5 = 32,500).

Regression

Model Summary - Income

Model	R	R²	Adjusted R²	RMSE
H₀	0.000	0.000	0.000	5.708
H₁	0.973	0.947	0.940	1.393

The model summary shows the influences of covariates (here, just one, Education) upon the dependent variable, Income. The statistics are the model fit measures for regression models in JASP.

The idea of model fit is how well the observed data fits with the model, in this case the idea that education affects income. The default model fit measures' for regression models in JASP are the coefficient R and what we would normally call the effect size, R squared. In regression, R squared is often known as the coefficient of determination, the proportion of variance in the dependent variable which is predictable by the independent variable(s). As usual, this ranges from 0 to 1. An effect of the magnitude of our fictional example is rare and unlikely in this particular scenario. The model summary table will assume greater importance when we test out different models using multiple regression, which is when we use Adjusted R squared and possibly RMSE.

ANOVA

Model		Sum of Squares	df	Mean Square	F	p
H₁	Regression	277.709	1	277.709	143.187	< .001
	Residual	15.516	8	1.939		
	Total	293.225	9			

Note. The intercept model is omitted, as no meaningful information can be shown.

The ANOVA table contains a range of reportable data. It can be reported in full, but of particular interest is the status of p. In this case, we can say that the regression is statistically significant.

Coefficients

Model		Unstandardized	Standard Error	Standardized	t	p
H₀	(Intercept)	29.450	1.805		16.316	< .001
H₁	(Intercept)	1.208	2.401		0.503	0.628
	Education	0.509	0.043	0.973	11.966	< .001

Chapter 7 – Tests of relationships

Regression

In the Coefficients table, the first thing to look at is the p value. As usual, we want a low p value to tell us that we have significance or, correctly, that we have evidence to support the rejection of the null hypothesis of no causal effect.

We then see how much of the dependent variable is predicted by each unit of the explanatory variable. The relevant statistic is the Unstandardized coefficient (also known as coefficient estimate, or just estimate). Here, you have to get your head around the idea that the figure referring to your predictor (or independent variable), here Education, actually refers to its influence on the dependent variable (Income). If we were dealing with single units, we would simply say that for every *one unit* increase in educational score, income would increase by .509 units (half a unit). However, our income is really in thousands, so we say that Income = 0.509 × 1000, giving $509. So every unit increase in educational grade is, on average, predictive of $509 in income.

Before leaving simple linear regression, let us consider the sales of subsidized burglar alarms to elderly people, working out the effects of prices on the number of sales at various shops. This example will be expanded when we cover multiple regression. The variable PriceA is the predictor (covariate); SalesA comprises the dependent variable. In general, the lower the prices, the higher the number of sales.

Model Summary - SalesA

Model	R	R²	Adjusted R²	RMSE
H₀	0.000	0.000	0.000	2292.732
H₁	0.553	0.306	0.270	1959.455

The proportion of variance is about 30%, using R squared, with p = 0.009. As the result is significant, we can consider the coefficient.

ANOVA

Model		Sum of Squares	df	Mean Square	F	p
H₁	Regression	3.218×10⁺⁷	1	3.218×10⁺⁷	8.382	0.009
	Residual	7.295×10⁺⁷	19	3.839×10⁺⁶		
	Total	1.051×10⁺⁸	20			

Note. The intercept model is omitted, as no meaningful information can be shown.

As well as p, the F ratio is important when it comes to comparing models.

Coefficients

Model		Unstandardized	Standard Error	Standardized	t	p
H_0	(Intercept)	8619.048	500.315		17.227	< .001
H_1	(Intercept)	15067.868	2268.105		6.643	< .001
	PriceA	-261.041	90.164	-0.553	-2.895	0.009

Note that the coefficient estimate for PriceA is negative, reflecting the inverse relationship between price and sales. It is always worth checking that the variables make sense; if they don't, it may be that a key variable has been left out of the model.

Before we consider other factors, it is worthwhile checking that the data are reliable. (Some of the data used here is not from a normal distribution as the author wished to achieve various effects with a small data set.) Let's visit Regression / Classical / Correlation and check out the plot for SalesA and PriceA.

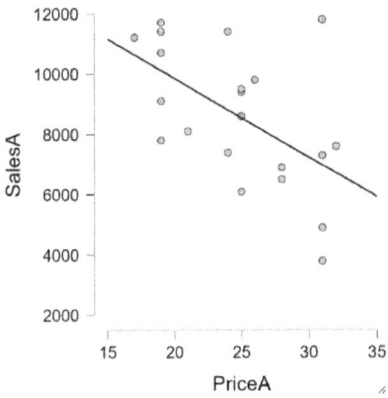

Note that we generally put the predictor (price) on the horizontal (x) axis and the dependent variable (sales) on the vertical (y) axis. Well, the relationship is linear, the bottom right to top left indicating a negative correlation, but there is a problem: an outlier on the top right of the chart (coordinates approximately 30, 12000). This is why it is advisable

Regression

to look at the data with a graph before focusing on tests. One coordinate is quite remote and is theoretically dubious: one shop is selling at the highest price and yet is also selling well. It could be a really exclusive shop, but let's not go there.

Usually in statistical testing, removing real information for convenience is unforgivable (unless they are input errors). However, we are building a predictive model. In such a case, it is acceptable to remove the outlier to improve the model. We want to study usual behavior.

Use the emended variables Sales and Price. These have omitted the outlying pairing (the last pair in the SalesA and PriceA variables, should you wish to examine the data).

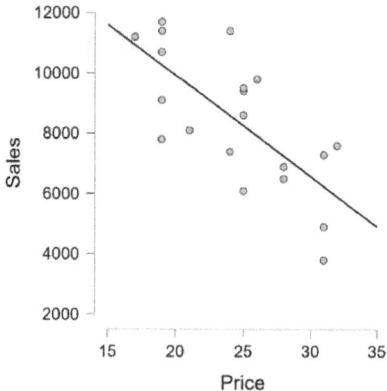

Note the more tightly knit pattern. Let us now update Linear Regression with the Dependent variable Sales and the predictor (covariate) Price.

Model Summary - Sales

Model	R	R²	Adjusted R²	RMSE
H₀	0.000	0.000	0.000	2230.270
H₁	0.715	0.511	0.484	1601.855

We now have R squared = 0.511, a big improvement. However, 0.511 still accounts for only half of the variance.

Regression

ANOVA

Model		Sum of Squares	df	Mean Square	F	p
H_1	Regression	$4.832 \times 10^{+7}$	1	$4.832 \times 10^{+7}$	18.832	< .001
	Residual	$4.619 \times 10^{+7}$	18	$2.566 \times 10^{+6}$		
	Total	$9.451 \times 10^{+7}$	19			

Note. The intercept model is omitted, as no meaningful information can be shown.

The larger the F ratio statistic, the greater the variation in group means. In this example, F is 18.832, a fairly substantial figure.

Coefficients

Model		Unstandardized	Standard Error	Standardized	t	p
H_0	(Intercept)	8460.000	498.704		16.964	< .001
H_1	(Intercept)	16628.499	1916.113		8.678	< .001
	Price	-334.912	77.177	-0.715	-4.340	< .001

Before we worry about that, let's do a little explanatory sum. As we are not dealing with thousands, but just single units, we can simply write that for every increase of a dollar in price, sales fall by, on average, about 335. Assuming that you want to escape from using the y and x beloved of many statistics books, you would write this formally as, Sales = +16628 -335 (Price). The first figure is the coefficient estimate for the Intercept, where the line of best fit passes through the y (dependent variable) axis.

It is likely, however, given the proportion of the variance, that additional variables may provide a more explanatory model. We will try this with multiple regression.

Multiple regression – multiple predictors against one dependent variable (parametric)

Returning to our example of sales from shops of subsidized burglar alarms to elderly people, we may ask if price is the only significant factor in determining the number of sales. Multiple regression allows us to build a model for effective prediction. We are particularly interested in two issues:

Regression

Will additional variables make an appreciable difference to predictions? If they do, are some variables more useful than others?

Press the Regression tab and select Classical / Linear Regression from the drop-down menu. Transfer Sales to the Dependent Variable slot, with the following variables to the Covariates box: Price, Instore, Street and Radio. As well as being interested in how far price affects sales, we are also interested in the contributory effects of promotion methods: are sales affected by promotions within the shops, street advertisements and on the radio? (Remember to use Sales and Price, not our rather less tightly knit variables SalesA and PriceA.)

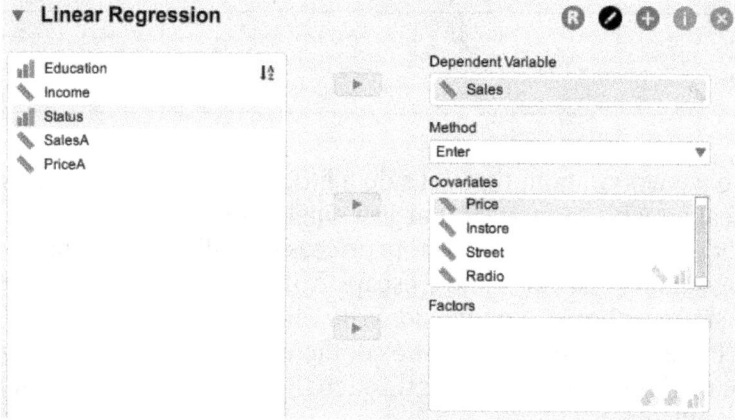

JASP also offers a Factors box. This means that you can also enter groupings, for example gender or class. You can have both covariates and factors at the same time, but do note that as usual, the more complicated the input, the more difficult it is to interpret the results.

Model Summary - Sales

Model	R	R²	Adjusted R²	RMSE
H₀	0.000	0.000	0.000	2230.270
H₁	0.878	0.770	0.709	1203.575

As our model reflects the complexity of the real world, the regression coefficient R is larger than the simple linear regression model of Price and Sales, with a much stronger R squared. However, Adjusted R squared is preferred with multiple regression; it takes into account the number of predictors and is more conservative than R squared. Our effect size is quite impressive: 0.709, about 70% of the variance.

ANOVA

Model		Sum of Squares	df	Mean Square	F	p
H_1	Regression	$7.278 \times 10^{+7}$	4	$1.819 \times 10^{+7}$	12.560	< .001
	Residual	$2.173 \times 10^{+7}$	15	$1.449 \times 10^{+6}$		
	Total	$9.451 \times 10^{+7}$	19			

Note. The intercept model is omitted, as no meaningful information can be shown.

The F test result is 12.560. This is substantively smaller than the statistic when Price was a lone predictor and suggests that not all of the variables are contributing much to the variance. We can be more specific when we look at the coefficients for the different variables:

Coefficients

Model		Unstandardized	Standard Error	Standardized	t	p
H_0	(Intercept)	8460.000	498.704		16.964	< .001
H_1	(Intercept)	-2105.259	4788.596		-0.440	0.666
	Price	-301.894	58.730	-0.645	-5.140	< .001
	Instore	1.384	1.235	0.152	1.121	0.280
	Street	0.472	0.348	0.182	1.356	0.195
	Radio	0.944	0.289	0.419	3.265	0.005

In addition to the inverse presence of Price, only Radio promotion appears to contribute significantly to the variance.

So our first question is answered positively in this instance: additional variables have made a considerable difference to the predictive model. It also seems likely that one of our additional variables, Radio, is more important than the others.

Before finding out about the extent of influence of the two main explanatory variables, it is sensible to check that we have met the statistical assumptions for multiple regression. Our predictions rely on the fact that the assumptions for the model hold.

As mentioned before, we need to check that we have normally distributed data and that the data is linear. We also need to check that

Regression

the errors between observed and predicted values – the residuals – are normally distributed. Yet another problem is too much multicollinearity, over-strong relationships between the variables. A further problem is autocorrelation, over-strong relationships between the residuals.

Let us first consider residuals. Within the Plots panel, choose 'Q-Q plot standardized residuals' and 'Residuals vs dependent'.

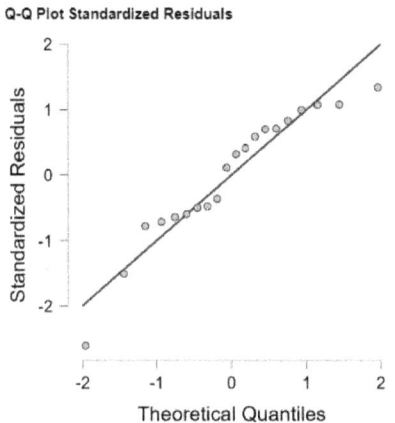

The residuals cling quite closely to the straight line, so we're happy with the Q-Q plot.

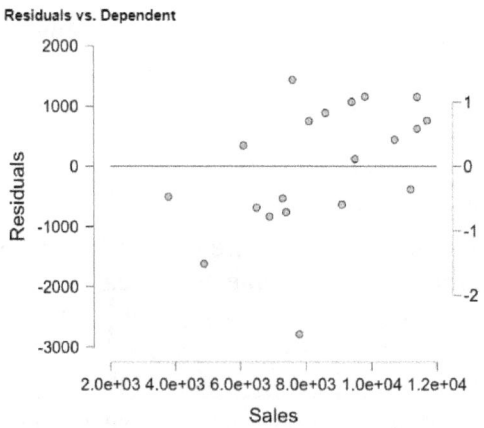

Regression

We are particularly interested in the dependent variable, Sales in this case. The residuals are quite randomly spread, which is fine. If the data formed a concentrated ball around the horizontal line, this also would not be a problem. What you do *not* want is either a U shape or its inverse, a rounded archway shape; these indicate curvilinear relationships.

Next we look at the collinearity and autocorrelation options. Within the Statistics panel, in the Coefficient section, choose Collinearity diagnostics, and in the Residuals section choose Durbin-Watson.

Coefficients

Model		Collinearity Statistics	
		Tolerance	VIF
H_0	(Intercept)		
H_1	(Intercept)		
	Price	0.975	1.026
	Instore	0.828	1.208
	Street	0.849	1.177
	Radio	0.929	1.077

High collinearity indicates over–correlated variables, which may mean that they are measuring the same underlying construct. As suggested elsewhere, throwing in a load of variables and hoping for something to come up is not a good idea; just include variables that make sense. We want a low VIF – variance inflation factor – not much above 1; these figures, between 1.03 and 1.21, are fine. We want a high Tolerance factor, approaching 1; the Tolerance column here is good, between .828 and .975.

Model Summary - Sales

Model	R	R²	Durbin-Watson		
			Autocorrelation	Statistic	p
H_0	0.000	0.000	0.643	0.713	< .001
H_1	0.878	0.770	−0.276	2.490	0.242

The Durbin-Watson statistic approaching 0 is a positive autocorrelation. If it approaches 4, there is a negative autocorrelation. The score should be between 1.5 and 2.5, as our result is, although some leniency could be shown as far as 1 and 3 respectively. However, it seems

Regression

reasonable to use the *p* value. The null hypothesis is that there is no autocorrelation between the residuals. The result (looking at H_1, the alternative hypothesis) does not support rejection of the null hypothesis; we're ok.

Another point to consider is sample size; results tend not to generalize with small samples. Stevens (1996) recommends 15 participants per predictor. Tabachnick and Fidell (2007) recommend the following formula: 50 participants + (8 x the number of predictors/covariates). So with 4 predictors, the minimum acceptable sample according to the former would be 4 x 15 = 60 participants, with the latter recommending 50 + (8 x 4) = 82. Given the small size of my fictional sample, it is unlikely that all of this would be replicated.

Returning to our results, we can run the multiple regression with only the predictors Price and Radio. This accounts for considerably more of the variance than our simple linear regression model using Price. The figures for the simple model (Price affecting Sales) were R = 0.715, R squared = 0.511, Adjusted R squared 0.484, F = 18.832. Here are the results of using the Price and Radio predictors.

Model Summary - Sales

Model	R	R²	Adjusted R²	RMSE
H₀	0.000	0.000	0.000	2230.270
H₁	0.833	0.694	0.658	1305.054

The model summary for Price and Radio gives us a considerably larger coefficient R and Adjusted R squared, the latter indicating two-thirds of the variance, compared to about half covered by the simple model. At the same time, there is a limited decline from the figures for the more comprehensive set of predictors (Adjusted R squared = 0.709). So, assuming that I have not omitted another useful predictor, we probably have the optimal model, explaining the sales figures via only two predictors while not drastically reducing the variance. The statistical term for explaining or predicting with as few explanatory variables as possible is a **parsimonious model**.

Regression

RMSE (root mean square error!) is an absolute measure of fit, which has as its advantage that it uses the same units as the dependent variable. Models with a smaller RMSE are indicated as having a better fit, but while RMSE tells you more about how well models fit the data, it is less easily interpretable than R squared. In our case, the RMSE is smaller for the comprehensive model than the model with two covariates, but we need to consider this proportionally: a drop of 101 when we are talking about figures of well over 1000 is not particularly impressive.

ANOVA

Model		Sum of Squares	df	Mean Square	F	p
H_1	Regression	$6.555 \times 10^{+7}$	2	$3.278 \times 10^{+7}$	19.245	< .001
	Residual	$2.895 \times 10^{+7}$	17	$1.703 \times 10^{+6}$		
	Total	$9.451 \times 10^{+7}$	19			

Note. The intercept model is omitted, as no meaningful information can be shown.

The F ratio is considerably higher for the model with two covariates than the comprehensive model. However, it is not much bigger than in the Price-only model.

Coefficients

Model		Unstandardized	Standard Error	Standardized	t	p
H_0	(Intercept)	8460.000	498.704		16.964	< .001
H_1	(Intercept)	3669.653	4362.780		0.841	0.412
	Price	-303.077	63.668	-0.647	-4.760	< .001
	Radio	0.973	0.306	0.432	3.181	0.005

Now we are in a position to estimate the effects of price and radio promotion on sales. As usual, ensure that the unstandardized coefficients make sense in relation to the model. As we are not dealing with thousands, as in our income example, but just raw numbers, we don't have to multiply anything. Sales decrease by 303, on average, for every dollar added to the price (note the negative coefficient) and there is almost one sale (.97) for each dollar spent on radio promotion. Formally, this can be written as
Sales = 3669.653 -303.077(Price) +0.973(Radio).

Regression

Bayesian equivalent

Open Regression / Bayesian / Linear Regression.

For this to work, it is recommended that all variables are considered 'Scale', showing the ruler symbol as above. To do this, click on the offending symbol, usually a set of upright bars representing 'Nominal', and choose Scale from the drop-down menu.

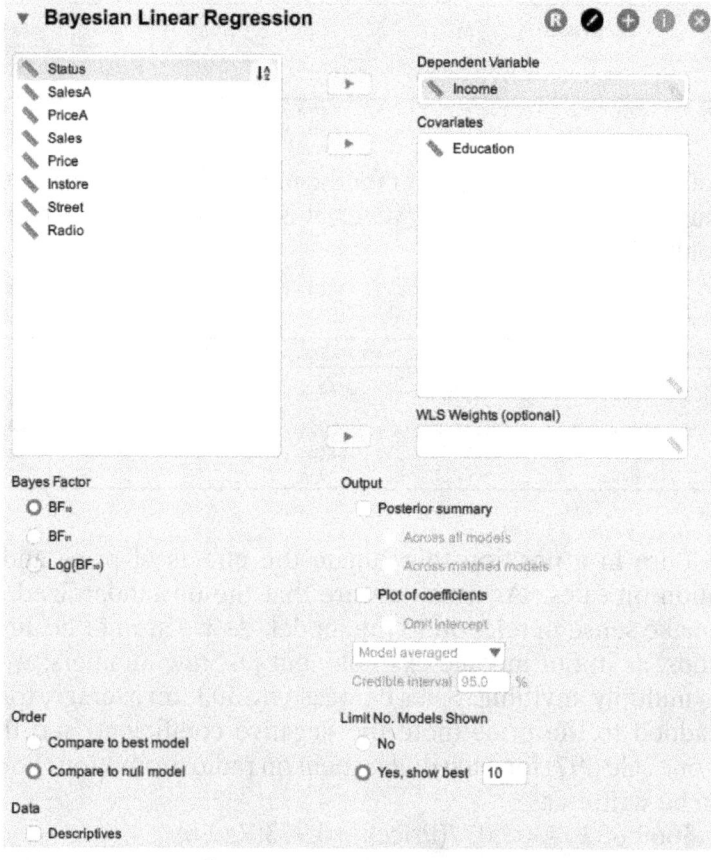

I have chosen 'Compare to null model' rather than the default setting. I keep things simple here, just concentrating on the Bayes factor, but if you want an explanation of 'Descriptives' and 'Posterior summary', consult a source such as Goss-Sampson (2020).

Model Comparison - Income

Models	P(M)	P(M|data)	BF_M	BF_{10}	R^2
Null model	0.500	2.479×10^{-4}	2.480×10^{-4}	1.000	0.000
Education	0.500	1.000	4032.642	4032.642	0.947

The Bayes factor is very strong indeed.

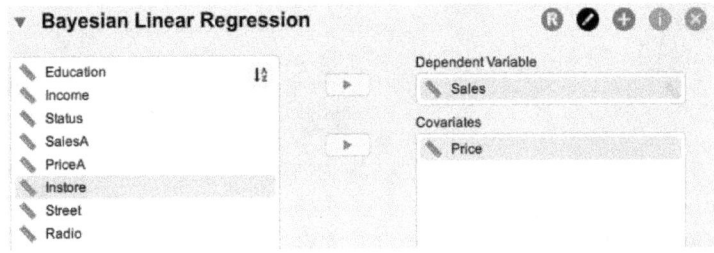

Model Comparison - Sales

Models	P(M)	P(M|data)	BF_M	BF_{10}	R^2
Null model	0.500	0.016	0.016	1.000	0.000
Price	0.500	0.984	63.398	63.398	0.511

The Bayes factor is within the 'Strong' banding (20–150).

Regression

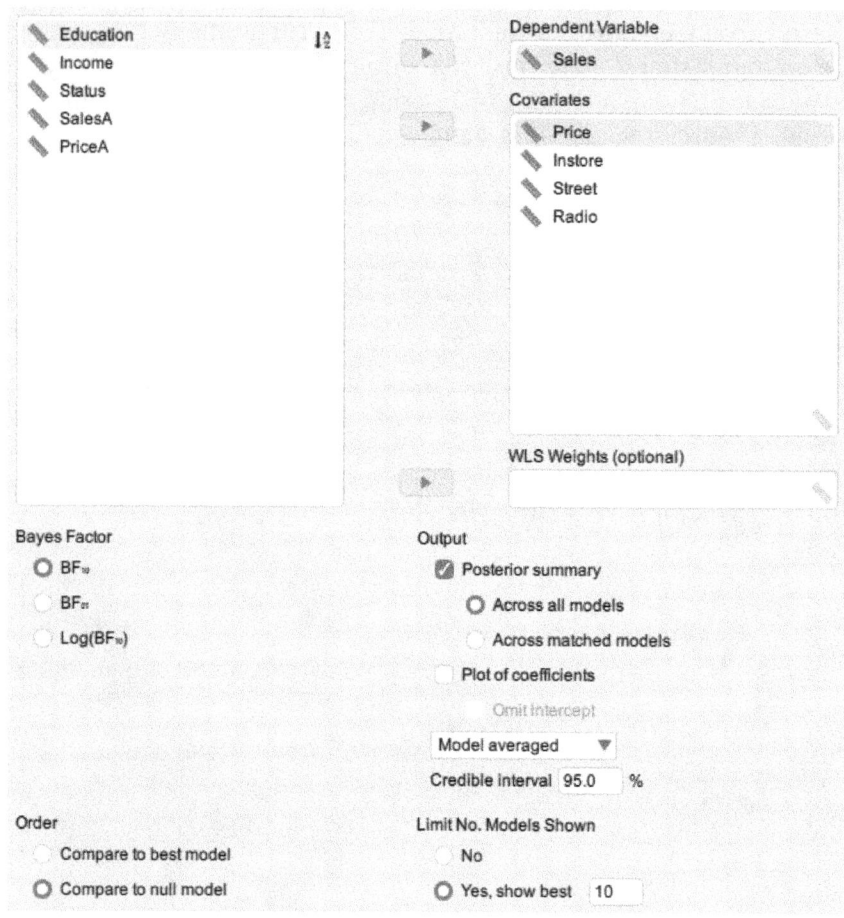

For multiple regression, I find a clear and simple usage for 'Posterior summary'. Note that I have left 'Limit No. Models Shown' on the default top 10 to save space; some situations may warrant the unlimited 'No'.

Regression

Model Comparison - Sales

Models	P(M)	P(M|data)	BF$_M$	BF$_{10}$	R²
Null model	0.200	0.001	0.006	1.000	0.000
Price + Instore + Street + Radio	0.200	0.420	2.895	288.888	0.770
Price + Street + Radio	0.050	0.217	5.254	596.140	0.751
Price + Instore + Radio	0.050	0.170	3.897	468.397	0.742
Price + Radio	0.033	0.129	4.289	531.867	0.694
Price	0.050	0.023	0.448	63.398	0.511
Price + Instore	0.033	0.017	0.499	69.832	0.594
Price + Instore + Street	0.050	0.010	0.200	28.676	0.607
Price + Street	0.033	0.009	0.258	36.402	0.555
Radio	0.050	0.001	0.026	3.828	0.285

Note. Table displays only a subset of models; to see all models, select "No" under "Limit No. Models Shown".

Posterior Summaries of Coefficients

Coefficient	P(incl)	P(excl)	P(incl|data)	P(excl|data)	BF$_{inclusion}$
Intercept	1.000	0.000	1.000	0.000	1.000
Price	0.500	0.500	0.995	0.005	189.113
Instore	0.500	0.500	0.619	0.381	1.625
Street	0.500	0.500	0.657	0.343	1.918
Radio	0.500	0.500	0.939	0.061	15.333

BF$_{inclusion}$ is brutally clear about which variables are important. *

Model Comparison - Sales

Models	P(M)	P(M|data)	BF$_M$	BF$_{10}$	R²
Null model	0.333	0.002	0.004	1.000	0.000
Price + Radio	0.333	0.939	30.733	531.867	0.694
Price	0.167	0.056	0.296	63.398	0.511
Radio	0.167	0.003	0.017	3.828	0.285

*This table has been truncated to save space.

Regression

Thinking point

We already know that too many variables tend to result in fluke correlations. Another weakness of correlations is that they do not prove cause and effect. For example, even if we are sure of a reliable relationship between confidence in the government and high levels of spending, are we sure that perceived stability leads to higher spending? Or, could it be that high disposable incomes predispose people to political complacency? Or should a mediating factor such as the unemployment rate be taken into account? Our use of regression, although assuming causation, does not avoid such uncertainties.

Running experiments or quasi-experiments is one way of dealing with this problem. When this is not practicable, we could triangulate: different aspects of a problem may be subjected to different forms of analysis, perhaps using different methodologies, to find out if the original theory can be disproven. Confirmatory factor analysis could be used to test the current model or to seek other explanatory models. Methods such as structural equation modeling can also examine the direction of correlational effects.

For those interested in the use of multiple regression in applied sociology, Bickel (2013) is highly recommended. This covers the sociology of education, primarily in West Virginia; the first section of the book, testing the theories of Karl Marx, is particularly readable.

This table refers to tests of relationships cited in this chapter. Each has a Bayesian equivalent in JASP, except for Spearman.

Purpose	Data	Number of variables	Test
Correlation	Parametric	2 or more	Pearson
Correlation	Non-parametric	2 or more	Spearman or Kendall's tau-b
Prediction and modeling	Parametric	2 or more predictors	Linear regression

Chapter 8 – Categorical analyses

Introduction

Speaking categorically

In this chapter, we are not interested in measurable data. We are interested in counts of observations, otherwise known as frequencies. In each example, the data fits into **exclusive and exhaustive** categories: each observation may only be included in just one category and the total number of categories must contain all of the observations in the study.

Let us say that your categories within a study of people with disabilities are Mental health problems, Learning disabilities and Physical disabilities. Each individual observation must go into just one of these categories; so exclusiveness means that you must make a decision as to whether a person with autism fits into the first or second of these categories, or in the case of a wheelchair-bound person with occasional depression, into the third or the first. In each case, the individual must only be placed in just one of the categories. Exhaustiveness means that you do not leave any individual outside of the categories.

In practice, this means that you have to make some subjective decisions. In the above case, you may need a further category, Multiple

Introduction

disabilities, for those with more than one primary disability. In other cases, you may wish to subsume categories into broader ones.

Angry	Irritated	Neutral	Positive
6	33	38	23

If you have this range of attitudes from a set of 100 interviews about a topic, you may wish to let a research commissioner know about the danger posed by the six individuals and hopefully the reasons. However, this type of statistical test will be heavily influenced by any discrepancy such as a really small (or large) number.

Given that the very small Angry category will certainly lead to a significant, and obvious, result, it would make sense in terms of the statistics to concatenate the Angry and Irritated categories into one category (Negative, for example). The decision to combine two or more categories is a subjective one.

This and the absence of measurement means that such statistics are often called **nominal**. A strong rationale needs to be balanced with statistical considerations. In the case of the above example, the combination of the two negative categories would make sense if you were interested in the proportions most likely to be disturbed by the issue.

The quantification of qualitative data

I often come across the assumption that qualitative research is something completely unrelated to quantification. Some of its proponents may even say outright that the results of their interviews or focus groups are not measurable. Without getting personal, I will suggest a few reasons why you would want to count, for example, the number of interviewees who think in a particular way about a subject or the number of focus groups changing their minds over a topic.

You might worry that the forceful viewpoint of a few respondents is clouding your judgement, and that many would not agree. You might also want to see if particular levels of status are more prone than others to share a particular view. Or you might want to look at the prevalence of a particular idea in comparison with conflicting ideas.

What's happening, statistically speaking?

At the core of these tests is a very simple principle: We are contrasting the **observed frequencies** with the **expected frequencies**. Are they significantly different?

The observed frequencies are the actual numbers that you have within each cell of the table, for example 23 Positive in the table above. The data in the cell comprises one category.

The expected frequencies are what should have been. In many cases, you will be uncertain of the likely result and therefore the expected frequencies are averages based on the numbers in each category. If, for example, you have 100 observations in 2 categories, then the random expected frequencies would be 50 in one and 50 in the other. In our chart, with 4 categories, the expected frequencies would be 25, 25, 25 and 25. That is why we do not really want disproportionately small or large observation frequencies in a category unless we really mean it, as the statistical test will go all weak at the knees and say, "McGinty, we've struck gold".

On other occasions, the expected frequencies are not random, but are based on what we have come to expect from previous results. Let us say that the incidence of racially motivated crime in a city has fairly predictable proportions against certain ethnic groups. Then a new study takes into account a relatively novel local phenomenon, perhaps immigration. In such a case, our statistical testing will examine whether the proportions are fairly comparable or different from the previously recorded results. So if our previous knowledge indicates expected proportions of .20, .30 and .50 within three ethnic categories of victims of crime, we would be interested to find out if the newly observed categories of offences have similar or different proportions.

The binomial test: a frequency test for dichotomies (either/or)

Quantification of categorical data is statistically simple but it does depend upon sound reasoning, taking into account logic within the research context.

The binomial test: a frequency test for dichotomies (either/or)

This test is for use with two categories only. Let us ask a sample of 30 individuals a simple question: "Do you think most people are honest – yes or no?" 13 people said 'yes' and 17 people said 'no' (we're in the realm of fiction as usual).

Open the **Frequencies1.csv** file. Note that it doesn't matter if data are in higgledy-piggledy order, as in the file. It also doesn't matter how you represent the levels; the first three columns from Frequencies1.csv are the same for the purposes of the binomial test, be it Yes or No, the typical 1 and 2 of dummy coding, or my idiosyncratic 16 and 7.

	Honesty1713Words	Honesty1713NumsA	Honesty1713NumsB
1	No	2	7
2	No	2	7
3	Yes	1	16
4	No	2	7
5	Yes	1	16
6	No	2	7
7	No	2	7

To use the test, press the Frequencies tab and select Classical / Binomial Test from the drop-down table.

The binomial test: a frequency test for dichotomies (either/or)

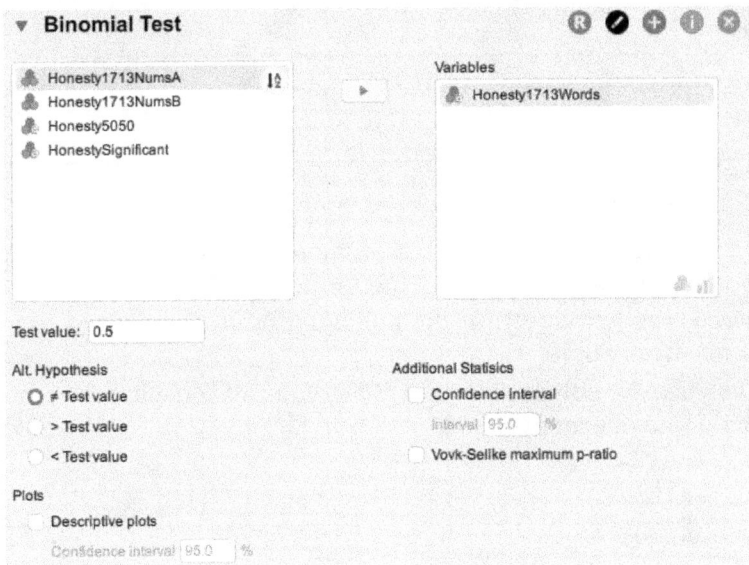

I have transferred the first column to the box on the right (any of the three '17/13' columns will behave similarly).

Note that the Test value default is 0.5, which means a random, 50/50 expectation. As there are 30 observations, the expected numbers would be 15 for each level (JASP's term for category or condition). It assumes either that we have no assumptions about proportions or that the results of previous studies were actually 50/50. The value can be adjusted to anywhere between 0 and 1. If we chose 0.75, for example, then previous studies would have had a ratio of three-quarters for one of the two conditions, a quarter for the other.

Note also the Alt. Hypothesis section. The default, that the results are different from the Test value, is the test's equivalent of a two-tailed test. It is rare for researchers to use one-tailed tests in this sort of situation. If there were a comparison of different conditions both of which were expected to have some effect, then a one-tailed hypothesis may be justifiable.

The binomial test: a frequency test for dichotomies (either/or)

Binomial Test

Variable	Level	Counts	Total	Proportion	p
Honesty1713Words	No	17	30	0.567	0.585
	Yes	13	30	0.433	0.585

Note. Proportions tested against value: 0.5.

As you can see, we get the counts of the level - always useful to check that your data entry is correct - the proportions (out of 1), and a high p value. There is support for the null hypothesis, that there is no clear difference from 50/50.

If you use the column Honesty5050, you can see what the test does with a sample of exactly 50/50.

Binomial Test

Variable	Level	Counts	Total	Proportion	p
Honesty5050	No	15	30	0.500	1.000
	Yes	15	30	0.500	1.000

Note. Proportions tested against value: 0.5.

This shows equal proportions and the highest possible value of p.

Let's now look at a significant result. Use the HonestySignificant column. Using the default settings, you will get this:

Binomial Test

Variable	Level	Counts	Total	Proportion	p
HonestySignificant	No	22	30	0.733	0.016
	Yes	8	30	0.267	0.016

Note. Proportions tested against value: 0.5.

We have a sample less impressed with this aspect of human nature. With a p value of 0.016, we can reject the null hypothesis of a similar number of responses on either side, unless of course we had specified a critical value of $p < .01$ or smaller. However, if we had good reason *before seeing the data* to expect a result in the 'No' direction, then we could select a

The binomial test: a frequency test for dichotomies (either/or)

one-tailed hypothesis (Test value); this gives a p value of 0.008, meeting the lower critical value.

But what if we had expected, from other studies, that our sample would be approximately 75/25? If you alter the Test value to .75 and click on the interface, you will then see a large p value against 'No'. In other words, we cannot reject the null hypothesis that No is similar to a three-quarters majority. Conversely, if you enter the value as .25, you will find the same p value against 'Yes', essentially a mirror image. So, the binomial test can test data against other proportions as well as 50/50.

Bayesian equivalent

Statistic		Quantification of evidence
Bayes Factor (BF10)	BF reciprocal (BF01)	
< 1	> 1	Noise
1 – 3	1 – 0.33	Weak
3 – 10	0.33 – 0.1	Moderate
10 – 20	0.1 – .05	Positive
20 – 150	.05 – .0067	Strong
> 150	< .0067	Very strong

This is a reminder of a suggested guide to Bayesian hypothesis reporting.

Using the **Frequencies1.csv** file, open Frequencies / Bayesian / Binomial Test.

The binomial test: a frequency test for dichotomies (either/or)

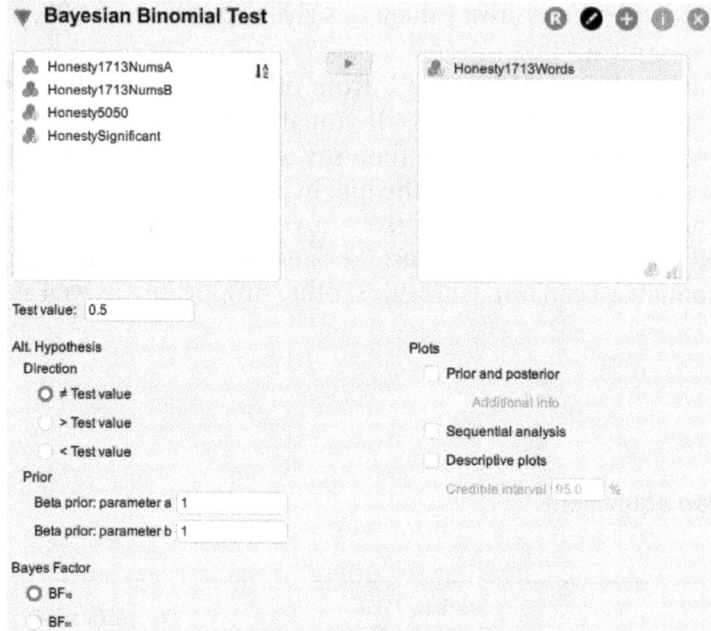

Data entry is the same as in the classical version of the binomial test.

Bayesian Binomial Test

	Level	Counts	Total	Proportion	BF_{10}
Honesty1713Words	No	17	30	0.567	0.289
	Yes	13	30	0.433	0.289

Note. Proportions tested against value: 0.5. The shape of the prior distribution under the alternative hypothesis is specified by Beta(1, 1).

The result, with an extremely small Bayes factor, provides no evidence in favour of the alternative hypothesis. If you select Prior and posterior / Additional info, you will find that the pizza chart is predominantly white, in favour of the null hypothesis (H0). A similar result will come from examining the variable Honesty5050.

Now consider the result of entering the variable HonestySignificant (remembering that the frequentist p value is .016):

Bayesian Binomial Test

	Level	Counts	Total	Proportion	BF$_{10}$
HonestySignificant	No	22	30	0.733	5.918
	Yes	8	30	0.267	5.918

Note. Proportions tested against value: 0.5. The shape of the prior distribution under the alternative hypothesis is specified by Beta(1, 1).

The Bayes factor is almost 6, within the Moderate band of credibility. The pizza will be largely dark red, in favour of the alternative hypothesis (H1). If we had opted for a one-tailed hypothesis, the Bayes factor would approach 12, within the Positive banding. As previously, you can alter to an expected Test value instead of .5. Try .90 or .10, for example.

The multinomial test: a frequency test for more than two categories

The multinomial test may be considered as an extension of the binomial test. It covers more than two conditions within one variable, and, like the binomial, can be used with even proportions as the default or can be set to compare the observed results with expected proportions. (The test is also known as the chi squared Goodness of Fit test. Please do not confuse it with what is normally referred to as 'Chi squared', with which you may be familiar, which covers two variables in 2 × 2 and larger grids. That is the chi squared test of Association, which appears after this test in Classical / Contingency Tables.)

Religious	Agnostic	Atheist
80	28	42

In this case, a sample of 150 people are asked for their religious views. We open **Frequencies2.csv** file to read in summary data of this data. (It is probably best to set up summary data for each multinomial test in a separate file; files with columns of different lengths tend to spook poor old multinomial and you can get some bewildering warning messages.)

The multinomial test: a frequency test for more than two categories

	Religiosity	Frequency
1	Religious	80
2	Agnostic	28
3	Atheist	42

To use the test, press the Frequencies tab and select Classical / Multinomial Test from the drop-down table.

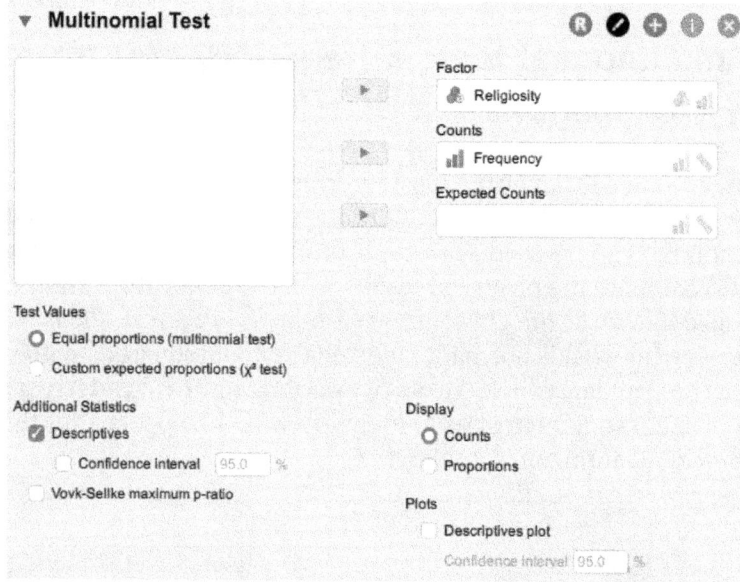

The titles of the categories go into the Factor slot and the corresponding frequencies go into the Counts box. (If using raw data, you would just use the Factor slot.)

The multinomial test: a frequency test for more than two categories

Multinomial Test

	χ²	df	p
Multinomial	28.960	2	< .001

Descriptives

Religiosity	Observed	Expected: Multinomial
Agnostic	28	50.000
Atheist	42	50.000
Religious	80	50.000

We have a large Chi squared statistic and the p value is a low one, below .001.

If you select the Proportions option in the Display section, instead of the default Counts option, it will be clear that each level was 'expected' to be equal; there are three equal slices (33.3% each) from our 'cake' of 150 responses. Our result tells us only that there is a significant difference from the expected, random, proportions. This does not say anything about any one category. If you wanted to isolate a specific condition, you would have to concatenate the categories and then run a binomial test for the remaining two categories, but the new categories would have to make sense. Perhaps in this example, you could divide the religious 80 from the sceptical 70.

For an example of a non-significant data set, let us consider a survey of 105 Londoners asked about the most stressful (note the exclusiveness) event likely to affect them, omitting bereavement.

Divorce	Unemployment	House-moving
35	37	33

From a new file, **Frequencies3.csv**, transfer the Stress column into the Factor slot and StressFreq into the Counts slot. A p value of 0.892 is a clearly non-significant result. (The expected counts are 35 apiece; our observed frequencies, as in the table, do not vary much from these.)

The multinomial test: a frequency test for more than two categories

As with the binomial test, this can be applied to proportions which are to be expected, often because of previous evidence.

Let us say that there are four major psychological disorders previously found in a recently integrated Amazonian community (this is fictional, as usual). We already have records of the relative prevalence of the disorders amongst known assimilated peoples; open the **Frequencies4.csv** file.

	Disorders	Frequency	Expected	ExpectedRatio
1	A	315	56.25	9
2	B	108	18.75	3
3	C	101	18.75	3
4	D	32	6.25	1

The Expected column represents percentages. So, in the known population, 56.25% of those with disorders suffered primarily from condition A and so on. Do note that the percentages for the whole sample should equal 100%. An alternative method, as shown on the ExpectedRatio column, is to use a ratio. If I had divided the types of condition into 16ths, I could have put in 9 for Condition A (9/16), 3 for Condition B, 3 for Condition C and 1 for Condition D. These produce the same results, so use whichever method suits you.

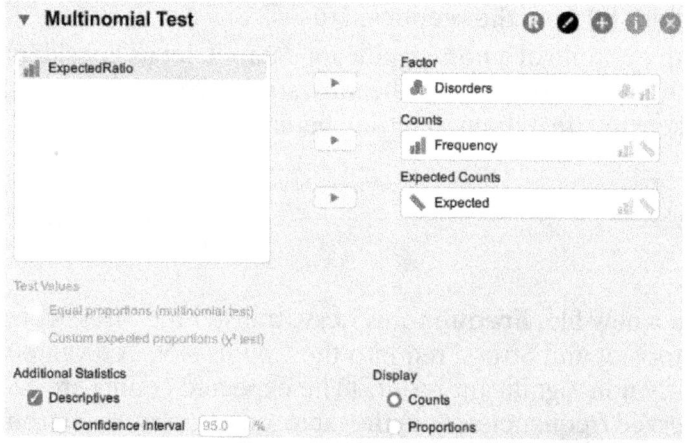

The multinomial test: a frequency test for more than two categories

As before, Disorders goes into the Factor slot and DisordersFreq into Counts. We already know the incidence of these conditions within known Amazonian peoples, so I place the Expected column (or ExpectedRatio) into the Expected Counts section. (If I hadn't put either of these columns into the file, I could have manually entered the expected values by going to the Test Values section and choosing Custom expected proportions.)

Multinomial Test

	x^2	df	p
Expected	0.470	3	0.925

Descriptives

Disorders	Observed	Expected: Expected
A	315	312.750
B	108	104.250
C	101	104.250
D	32	34.750

In the results, I show the 'Descriptives' statistics from the Additional Statistics section. The Expected counts show a close correspondence to the new sample (Observed). The incidence of conditions appears to be similar to those of the known population. Returning to the technicalities, we see a small Chi squared statistic and a high p value, indicating a non-significant result: the observed frequencies are fairly similar to the expected frequencies.

Bayesian equivalent

Using the **Frequencies2.csv** file, open Frequencies / Bayesian / Multinomial Test.

The multinomial test: a frequency test for more than two categories

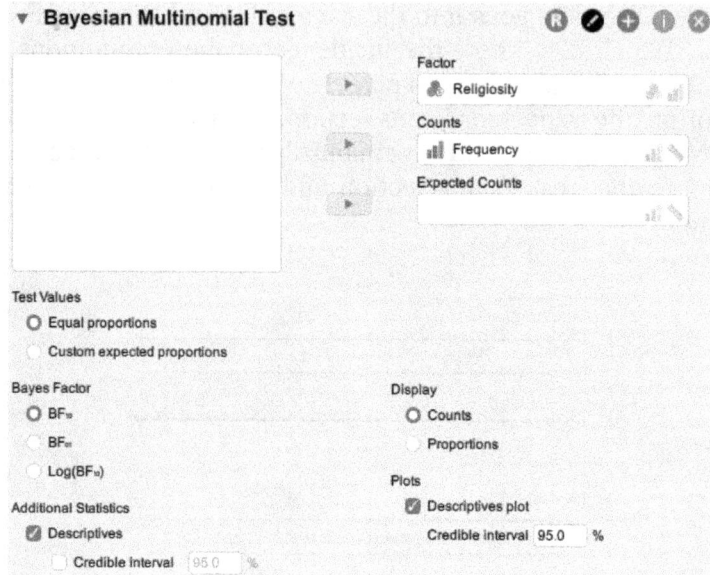

As usual, the data entry is the same as in the classical version.

Bayesian Multinomial Test

	Levels	BF$_{10}$
Multinomial	3	17300.196

The Bayes factor is huge, in line with the classical result. Some would say that the alternative hypothesis is over 17,000 times more likely than the null hypothesis; I'm not so sure about such a quantification.

If you open the **Frequencies3.csv** file, previously shown to have a non-significant result, the Bayes factor is very small.

The above results are based on 'expected' counts of equal proportions. Let's open **Frequencies4.csv** to customize the expected proportions:

The chi squared Test of Association: a frequency test for two variables

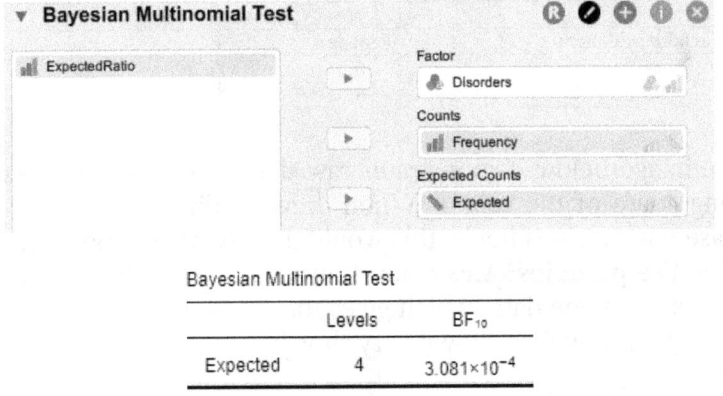

Bayesian Multinomial Test		
	Levels	BF_{10}
Expected	4	3.081×10^{-4}

The negative scientific notation represents an extremely small Bayes factor. The current and expected proportions do not differ.

The chi squared Test of Association: a frequency test for two variables

Often known as just **Chi Squared**, probably because most introductory textbooks teach only this categorical test, the chi squared Test of Association is used to find out whether or not there is a relationship between variables. The test is also known as the chi squared Test of Independence, perhaps a purer statistical definition, as the null hypothesis is that the variables will be independent of each other. We may want a 'significant' relationship; the computer seeks a lack of significance, the null hypothesis.

The contingency table allows us to study two or more variables in tandem. The statistical test itself, chi squared, distinguishes whether or not there is a relationship between the variables.

Before going into this in more detail, it is worthwhile knowing that there are two ways of entering the data. First, let us look at *raw data*. Here we use two variables, Ethnicity and Gender, each with two conditions/levels:

The chi squared Test of Association: a frequency test for two variables

	Variable: Gender	
Variable: Ethnicity	Female	Male
Non-White	6	6
White	6	12

The image below demonstrates raw data from **Frequencies5.csv**, showing some of the cases. When drawing directly from a suitable database with various fields, this would usually be appropriate. (If you examine **Frequencies5A.csv**, not shown here, you will find a summary version of the same data, which gives the same statistical results. This method of input will be shown very shortly.)

Ethnicity	Gender
Non-white	Male
Non-white	Male
Non-white	Female
Non-white	Female
White	Male
White	Male
White	Male
White	Male
White	Female
White	Female

For this test, press the Frequencies tab and select Classical / Contingency Tables from the drop-down menu.

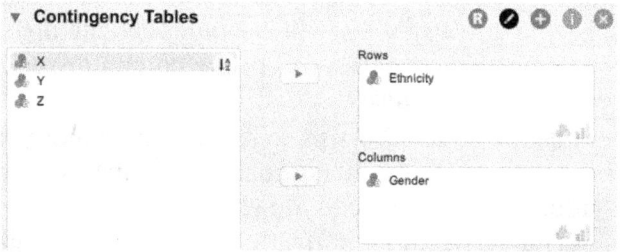

All you need to do is to transfer one variable into the Rows box and the other into the Columns box. This is only a 2 × 2 grid, White and

The chi squared Test of Association: a frequency test for two variables

Non-white versus Male and Female, so the order of rows and columns does not matter much. Where a variable has many levels, it may be a good idea to put the variable with more conditions into Rows for neater output.

The output shows a large *p* value, 0.361. We cannot reject the null hypothesis, that there is no significant relationship between the two variables Gender and Ethnicity.

Moving on, let us extend our interest in religion, as when looking at the multinomial test, to examining the views of citizens of two different countries. The chart shows the design, a grid of 2 × 3 (levels/conditions), although still comprising only two variables, Religiosity and Country.

	Variable: Gender	
Variable: Religiosity	Males	Females
Religious	46	42
Agnostic	11	15
Atheist	23	17

Note the exhaustiveness and exclusivity rule: We have 154 people in our sample, all are allocated, each being allocated to just one of these cells.

Here we use *summary data*, from **Frequencies6.csv**. The first two columns represent the two variables; between the two, we cover all of the permutations of levels across the two variables. There then follows the frequency for each cell.

	Gender	Religiosity	Frequency
1	Males	Religious	46
2	Males	Agnostic	11
3	Males	Atheist	23
4	Females	Religious	42
5	Females	Agnostic	15
6	Females	Atheist	17

The reason I use a different file for each summary table is that the contingency table's Counts field gets confused by different column lengths.

Chapter 8 – Categorical analyses

The chi squared Test of Association: a frequency test for two variables

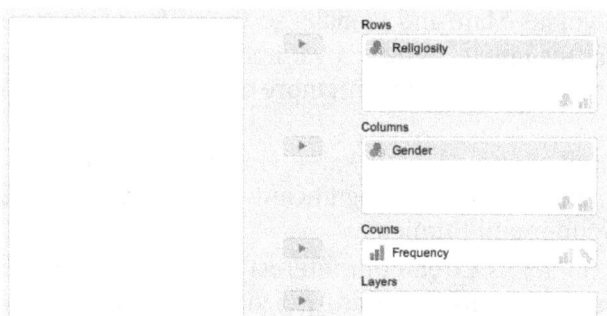

We transfer the two summary variables into the Rows and Columns boxes, placing the frequencies variable in the Counts box. In the output, you will see a *p* value of 0.481; the null hypothesis may not be rejected.

Give me a significant result, I think I hear you say.

We might be considering the effects of a television documentary, perhaps on immigration, or domestic violence, or perhaps studying what we believe to be a moral panic. Let us consider asking two questions: Have respondents seen or heard about an issue? Are they concerned about an issue? Do note that the test only allows us to consider whether or not there is a relationship between the two variables; it cannot ascertain causal direction.

		Concern	
		Serene	Worried
Awareness	Not Seen	30	12
	Seen	12	18

	Awareness	Concern	Frequency
1	Seen	Worried	18
2	Seen	Serene	12
3	Not seen	Worried	12
4	Not seen	Serene	30

The data, as in the file **Frequencies7.csv**, contains pairings of the different levels (2 × 2) across the two variables, Awareness and Concern. Each pairing has a frequency count, the number of observations per category. So we have 72 individuals, each allocated to just one cell.

The chi squared Test of Association: a frequency test for two variables

The variables are placed in Rows, Columns and Counts, as in the previous example.

Contingency Tables

	Concern		
Awareness	Serene	Worried	Total
Not seen	30	12	42
Seen	12	18	30
Total	42	30	72

Chi-Squared Tests

	Value	df	p
X^2	7.112	1	0.008
N	72		

We have a small p value, 0.008, which would meet the critical value of $p < .01$.

A note on samples: It is generally recommended that there should be at least 20 observations per sample, with at least 5 observations per cell. On the other hand, where you do have some cells with 5 or less (some say below 10), you may choose to go into the Statistics panel and choose the Chi squared continuity correction (also known as the Yates Correction). Actually designed for small numbers of expected cells, this adjustment gives a more conservative assessment, with a higher p value. However, it has a tendency to produce Type 2 errors (wrongly supporting the acceptance of the null hypothesis or, in layman's terms, falsely indicating non-significance). Many statisticians believe that it should never be used. It is certainly not necessary in this case.

To complicate your life further, you should know that very large samples often lead to very small p values. In fact, a large sample may appear to be 'more significant' than a small sample with the same relationship between observed and expected values. Here, measures of effect size, the magnitude of the relationships, really come into their own. They adjust for sample size and are thus more usable for larger samples than chi squared and the p value.

The chi squared Test of Association: a frequency test for two variables

The most commonly reported of these 'measures of association', as they are also called, are Phi and Cramer's V, to be found in the Nominal section of the Statistics panel. Both have the useful attributes of running from 0 to 1, from no relationship to being exactly the same. Phi is usually read for 2 x 2 tables, with Cramer's V for larger contingency tables (for example, one variable is comprised of children, teenagers and adults, another of males, females and 'other gender'.)

In the promotional campaign example, a 2 x 2 table, both Phi and Cramer's V give 0.314 as the magnitude of the effect. Cohen (1988) provides a rule of thumb in which 0.1 is a small effect, 0.3 a medium effect and 0.5 a strong effect. His examples of their comparative magnitude are the difference in mean height between 15- and 16-year-old girls (small), the difference between 14- and 18-year-old girls (medium) and that between 13- and 18-year-old girls (large).

With larger contingency tables, such as a 3 x 3, smaller effect size statistics emerge. To interpret them, the df statistic displayed in the output next to Chi squared and the p value become of use. The degrees of freedom (df) refer to the number of factors free to vary in the calculations. The df statistic is calculated by taking the combined number of rows and columns and subtracting 1.

df	small	Medium	large
1	.10	.30	.50
2	.07	.21	.35
3	.06	.17	.29
4	.05	.15	.25
5	.04	.13	.22

This table is adapted from Cohen (1988). As can be seen, smaller Cramer's V statistics are accepted when larger grids are in use.

One question which may occur to you is what should you do when you come across figures such as .2 and .4? Personally, I would consider .4 a strong effect size; indeed, effect sizes of more than .5 are to be

The chi squared Test of Association: a frequency test for two variables

viewed with some suspicion, as they may indicate that the different variables are measuring similar concepts.

In general, however, it should be noted that the above recommendations are a rule of thumb. Knowledge of previous studies covering similar ground are generally considered to be more helpful – if you can find them. If in doubt, just cite the statistic without committing to a description of the size.

I have not consulted the ordinal tests, Gamma and Kendall's *tau-b*, in the Ordinal section, as our variables are not gradable in terms of magnitude. A possible ordinal example would be responses to a video about violence against women: one side of the grid would be physical signs of anger, angry verbal responses not including violence, and facial signs of irritation not including violence or angry verbal responses, the other side could comprise violent criminals, non-violent criminals and members of the public. This assumes that we agree that both of these variables are gradable, in terms of levels of reaction and of antisocial tendency.

Gamma should be interpreted as follows: .75 to 1 = a Strong relationship; .5 to .74 = Moderate; .25 to .49 = Weak; < .25 = No relationship. If you look at Gamma in the (inappropriate) promotional campaign example, you will find that the result is just about within the Moderate reporting category: essentially, tests for ordinal data, that is with categories of increasing magnitude, are generally more demanding than those merely seeking a relationship without such a direction. *Tau-b* may be better for grids with the same number of rows and columns in the table.

Remembering that the test result only comments on whether or not the overall relationship between the two variables is significant (or rather, allows us to reject the null hypothesis), it is often helpful to take a closer look at the grids. The Cells panel provides a wealth of information about what has been going on inside the test. First, within Counts, select Expected to go alongside the default Observed counts.

The chi squared Test of Association: a frequency test for two variables

Contingency Tables

Awareness		Concern		Total
		Serene	Worried	
Not seen	Count	30.000	12.000	42.000
	Expected count	24.500	17.500	42.000
Seen	Count	12.000	18.000	30.000
	Expected count	17.500	12.500	30.000
Total	Count	42.000	30.000	72.000
	Expected count	42.000	30.000	72.000

In this particular case, all of the cells have a strong difference between the observed count and the expected count. This is not always the case. In this particular instance, we can see that those who have seen the promotion are less likely to be serene and more likely to be worried than if the numbers were randomly distributed (the null hypothesis). The opposite is the case for those who have not seen the promotion.

Do not assume causality, however. It could be that seeing the promotion increases anxiety about the issue, but it is also possible that being anxious about an issue makes people more likely to look for materials which reflect their feelings. Another possibility is a mediating factor: perhaps additional news coverage could increase both anxiety and the tendency to follow the promotion.

In cases where things are not certain, it is possible to reduce categories in order to subject them to the multinomial test or the binomial. However, you would need to be sure that the concatenated categories make good theoretical sense or at least form a coherent rationale. The usual rules of exhaustiveness and exclusivity apply. Here, we strip away the Expected counts and just look at the default Observed counts.

Contingency Tables

Awareness	Concern		Total
	Serene	Worried	
Not seen	30	12	42
Seen	12	18	30
Total	42	30	72

The chi squared Test of Association: a frequency test for two variables

In this case, you would use the outer figures to look at single variables. If examining Awareness, you would look to the far right, using the 42 and 30 for a binomial test. You would do likewise for the figures at the bottom for the Concern variable. With a larger grid than this, the proportions would be less simple and you would use the multinomial test, where a variable has more than two conditions.

Another thing that you can do is to study the Percentages within the Cells panel. I find these particularly useful for reporting results. Personally, I would look at them one by one. If you have a particular area of interest, you may decide to view only one set of these.

Contingency Tables

Awareness		Concern		Total
		Serene	Worried	
Not seen	Count	30.000	12.000	42.000
	% of total	41.667 %	16.667 %	58.333 %
Seen	Count	12.000	18.000	30.000
	% of total	16.667 %	25.000 %	41.667 %
Total	Count	42.000	30.000	72.000
	% of total	58.333 %	41.667 %	100.000 %

Total: The outer figures show the overall totals per variable. So if you wanted to report on the percentages for the Awareness variable you would look at the right and use Not Seen 58.3% and Seen 41.7% (58% and 42% respectively if rounding the figures further). The same is true for the Concern variable, using the figures at the bottom. As before, these would vary more when you deal with a different-shaped grid. Inside, you have the possibility of reporting any of the categories in terms of percentages (for example, those who had seen the promotion and were worried comprised 25% of the sample). Often, it is a good idea to report percentages, as they usually demonstrate proportionality more effectively than absolute numbers.

The chi squared Test of Association: a frequency test for two variables

Awareness		Concern		
		Serene	Worried	Total
Not seen	Count	30.000	12.000	42.000
	% within row	71.429 %	28.571 %	100.000 %
Seen	Count	12.000	18.000	30.000
	% within row	40.000 %	60.000 %	100.000 %
Total	Count	42.000	30.000	72.000
	% within row	58.333 %	41.667 %	100.000 %

Row: This is where you are interested in the relationship between the figures *within the row variable*. To give this some flesh, imagine that we have replaced the 'Not seen' row with Females and 'Seen' with Males. You would then be able to give figures for Females: "Of the females in the sample, a little over 70% were reasonably relaxed about the issue." For Males: "Of the males, 60% were worried."

Awareness		Concern		
		Serene	Worried	Total
Not seen	Count	30.000	12.000	42.000
	% within column	71.429 %	40.000 %	58.333 %
Seen	Count	12.000	18.000	30.000
	% within column	28.571 %	60.000 %	41.667 %
Total	Count	42.000	30.000	72.000
	% within column	100.000 %	100.000 %	100.000 %

Column: Here, we view the proportions from the top down. Let us do a similar trick of the imagination by replacing 'Serene' with Left-wingers and 'Worried' with Right-wingers (apologies to the center ground). 71% of the Left have not seen the promotion. 60% of the Right have seen it. (As usual, we are still left with whether or not right-wingers were more sensitized to the issue and therefore more likely to see the promotion, or perhaps the promotion somehow appeared in places or media more likely to be seen by right-wingers, or some other mediating factor.)

You will also find the Log Odds Ratio (for two by two variables only), within the Statistics panel. This provides a measured way of looking

The chi squared Test of Association: a frequency test for two variables

at the strength between two effects, moving beyond the traditional hypothesis testing method of Chi-squared. A log odds ratio of 1 means that the two effects are independent of each other; there is no interactive effect. A ratio greater than 1 – in our example, you should see the figure 1.322 – indicates that there is some interdependence. If the result had been less than 1, then one effect would have been evident, but there would be no interrelationship (Swan, 2021).

A note on the Layers box. This is for additional variables. Here, we split the data we had on promotion between males and females (to be found in the **Frequencies8.csv** file):

	Aware	Concern	Gender	Frequency
1	seen	worried	female	10
2	seen	serene	female	7
3	not seen	worried	female	4
4	not seen	serene	female	16
5	seen	worried	male	8
6	seen	serene	male	5
7	not seen	worried	male	8
8	not seen	serene	male	14

If the 'Gender' variable is placed in the Layers box, we will get the same overall results as previously, but also a separate examination of each of the gender levels. Beware of using more than one layer variable at any one time, as the results may be very difficult to interpret.

Bayesian equivalent

Using either **Frequencies5.csv** or **Frequencies5A.csv**, open the menu: Frequencies / Bayesian / Contingency Tables. This involves Ethnicity and Gender, and the data input is completely the same as with the Classical version of the tables.

The chi squared Test of Association: a frequency test for two variables

Bayesian Contingency Tables Tests

	Value
BF_{10} Independent multinomial	0.633
N	30

Note. For all tests, the alternative hypothesis specifies that group Non-white is not equal to White.

The classical test result was a large p value. It will come as no surprise to see a Bayes factor of less than 1, essentially noise. (Note that the rubric about gender would change to ethnicity if you exchanged the contents of the rows with the columns.)

If you open **Frequencies6.csv** and enter the variables Religiosity (more than two conditions), Gender, and Frequency, you will find, in line with the non-significant frequentist result, a small Bayes factor.

Seeking a significant result, previously shown as p = 0.008, we open **Frequencies7.csv**, entering Awareness, Concern, and Frequency:

Bayesian Contingency Tables Tests

	Value
BF_{10} Independent multinomial	9.408
N	72

Note. For all tests, the alternative hypothesis specifies that group Not seen is not equal to Seen.

The result does not appear to be quite as enthusiastic as the classical one: 9.408 comes near the top of the Positive credibility banding.

Opening **Frequencies8.csv**, we can enter Aware, Concern, Frequency, and Gender, the last of these is placed in the layers dialog box. We see an overall result, BF10 = 9.408, and results according to gender:

Bayesian Contingency Tables Tests

Gender		Value
female	BF_{10} Independent multinomial	6.516
	N	37
male	BF_{10} Independent multinomial	1.103
	N	35
Total	BF_{10} Independent multinomial	9.408
	N	72

Note. For all tests, the alternative hypothesis specifies that group not seen is not equal to seen.

Log-linear regression: modeling three or more categorical variables

If you have more than two categorical variables, log-linear regression starts with a large model representing the data and looks for more parsimonious smaller models. As the name indicates, this technique uses regression techniques, so if you haven't already read about regression, it might be sensible to do so in order to enhance your understanding of the output.

Let us consider a (fictional) study of religiosity and political leanings across three countries.

		Religiosity		
Country	Politics	Religious	Agnostic	Atheist
Denmark	Conservative	38	16	26
	Progressive	14	32	24
Germany	Conservative	46	11	24
	Progressive	8	34	27
Chile	Conservative	63	13	12
	Progressive	19	27	16

From our previous research, we expect Chileans to be the most religious nation, the Danes to be, on average, the least religious and the Germans to come somewhere in between. We also have reason to believe that people with conservative views are more likely to be religious than people with more progressive politics. What we don't know is whether or not this will translate into the citizens of the different countries tending to have different political leanings.

Log-linear regression: modeling three or more categorical variables

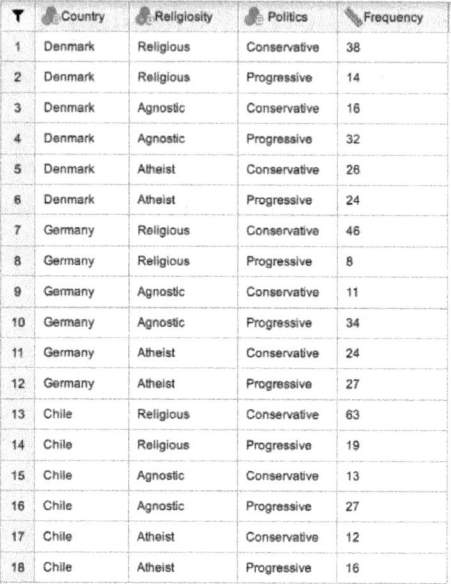

In the file **Frequencies9.csv** each permutation of levels has its own row, including a count cell.

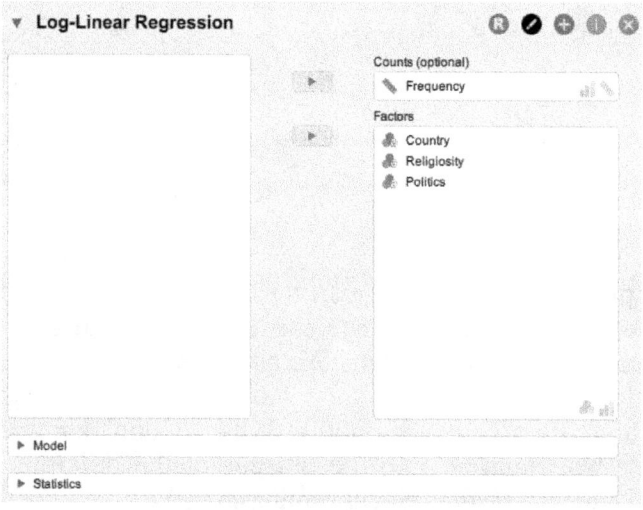

Log-linear regression: modeling three or more categorical variables

From the menu, press the Frequencies tab and select Classical / Log-Linear Regression.

ANOVA

	df	Deviance	Residual df	Residual Deviance	p
NULL			17	121.382	
Country	2	1.421×10^{-14}	15	121.382	1.000
Religiosity	2	13.995	13	107.387	< .001
Politics	1	5.130	12	102.257	0.024
Country * Religiosity	4	17.823	8	84.434	0.001
Country * Politics	2	1.028	6	83.406	0.598
Religiosity * Politics	2	79.310	4	4.096	< .001
Country * Religiosity * Politics	4	4.096	0	0.000	0.393

Deviance refers to the deviance of the fitted logistic model from a perfect model; higher values demonstrate larger effects. The p values show that there is little evidence to support rejection of the null hypothesis for the relationship between countries and politics, that there is no relationship between them. This and, unsurprisingly, the higher level model of an interaction between countries, religiosity and politics, can be discounted. The relationships between religiosity and politics, and between country and religiosity may be worthy of further analysis.

Where next? The simplest way is to collapse the study into interactions of two effects at a time, using the chi-squared test of association (Classical / Contingency Tables) as in the previous section of this chapter. This offers in-depth analyses, which could include the 'observed' and 'expected' observations, and percentages.

When you have a lot of variables, probably using raw data, using log-linear analysis as a filtering system may make a lot of sense.

Bayesian equivalent

Opening **Frequencies9.csv** and using Frequencies / Bayesian / Log-Linear Regression, you will find that the data entry is exactly the same as in the frequentist version. Enter the Frequency variable in the Counts slot; inside the Factors box, place Country, Religiosity, and Politics.

Log-linear regression: modeling three or more categorical variables

Model Comparison

	Models	P(M|data)	BF$_{10}$
1	Country + Religiosity + Politics + Country * Religiosity + Religiosity * Politics	0.892	1.000
2	Country + Religiosity + Politics + Country * Religiosity + Country * Politics + Religiosity * Politics	0.077	0.086

Note. Total number of models visited = 5

Looking at the somewhat oblique output, it is probably best to discard the model with the small Bayes factor, the model referred to as 2; this contains all of the three factors and all of the three interactions and is a 'saturated model' (covers all the bases, but I mean that in a negative way). The model referred to as 1 is notable for what is omitted: it does *not* include the interaction between Country and Politics ('Country * Politics'), which was also discounted by the frequentist test.

You will notice that the BF$_{10}$ read-out on the table above will vary a little each time the test is run. For somewhat more consistent results, you may choose to open the Advanced panel, and switch from the default 'Auto' to 'Manual'; I would suggest leaving the 10000 figure, although it will take some time to process:

```
▼ Advanced

Samples              Repeatability
  ○ Auto               □ Set seed: 1
  ● Manual
    No. samples 10000
```

Number of variables	Test
1 – with two conditions	binomial
1 – more than two conditions	multinomial aka chi squared Goodness of Fit
2 – various conditions plus possible layers	contingency tables – **chi squared** aka chi squared Test of Association / or chi squared Test of Independence
3 or more	log-linear regression

All of these tests have a Bayesian equivalent in JASP.

Chapter 9 – Exercises

These exercises allow you to test your learning of Chapters 6, 7 and 8.

Questions

Question 1

Members of a rehabilitation group for schizophrenic outpatients decide between two strategies for dealing with persistent unwelcome thoughts (no 'abstaining'). Of 40 opinions offered, one of the policies was chosen by 26 people. What test should be used and what is the outcome? Accept a significance level of $p < .05$

Question 2

On a rating scale of difficulty, are male clinical psychologists more likely than females to see their job as stressful? Psychologists of both genders were given the same five point rating scale to complete. What test is appropriate?

Question 3

There is a proven relationship between scores on family dysfunction scales and disruptive incidents at school. You have a range of variables such as types of behavior, age of siblings and family income. What method would be most appropriate?

Questions

Question 4

In a sample of industrial psychologists, a study is being undertaken of the settings in which they work.

	Careers Guidance	Business	Government	Academic
Female	6	11	20	30
Male	8	18	16	32

What method should be used? Are there significant differences?

Question 5

Supporters of different political parties are asked which of three explanations they prefer for why individuals behave well.

Belief A = family upbringing; Belief B = the social milieu; Belief C = the educational system.

Of the Red Party, 100 favored Belief A, 230 favored Belief B and 150 favored Belief C. The Blue Party favored Beliefs A, B and C as follows: 600, 400, 100. The Yellow Party favored Beliefs, A, B and C as follows: 180, 120, 100.

Consider an appropriate design and find out if the test results are significant.

Question 6

A correlation matrix containing a lot of variables includes many correlations at .9   What should you do?

Question 7

Can a correlation coefficient of .2 be significant?

Answers

Answer 1

This 'yes or no' situation, a dichotomy, can be subjected to the binomial test.

If we had a clear and well-reasoned rationale about one policy being superior to the other, prior to examining the data, then a one-tailed hypothesis could have been used. The binomial test would have supported rejection of the null hypothesis. Unless this was the case, the two-tailed hypothesis needed to be chosen: the result would not be significantly different from chance.

Answer 2

Mann-Whitney examines the differences between two different sets of individuals. (If the rating scales had been calibrated, the independent samples ('unpaired') t test could have been used.)

Answer 3

Multiple regression. You would want to see how well different variables fitted into a model.

Answer 4

Chi squared Test of Association. The result is non-significant.

If you collated the different settings categories, ignoring gender by adding the two groups together, and subject them to the multinomial (Goodness of Fit) test, you would see a clearly significant result.

You could also use the binomial test to study the two genders, ignoring settings. As the number of men and women in this sample are quite similar, this would be a non-significant result.

Answers

Answer 5

	family	social	education
Red	100	230	150
Blue	600	400	100
Yellow	180	120	100

Chi squared Test of Association is the appropriate test, with a clearly significant result.

If you look at the expected and observed counts, you will see that Blues in this sample are much more likely to believe in the influence of the family on behavior and less likely to see this as educational. Reds are in particular less likely to see family as influential.

Answer 6

The problem is likely to be collinearity. Within (classical) linear regression, go to the Statistics panel and use 'Collinearity diagnostics'. Some of your variables probably have very similar meanings and should be removed. Similarities are useful, but not duplication. When your cull has lowered the collinearity of your data set, you could then use principal components analysis (see Chapter 12) to reduce data further, allowing a more in-depth investigation of the correlations.

Answer 7

This sort of correlation coefficient can be quite common when dealing with large data sets and may be accompanied by an acceptable p value. If you are interested in the magnitude of the effect, perhaps for applied usage, you may wish to consider the effect size, .04. In some contexts, 4% of the variance may be important; in other cases, such an effect size is negligible.

Chapter 10 – Reporting research

As each university, and often each department, has its own guidelines for reporting, students are urged to consult the relevant guidelines. The basics, however, are the test statistic, the p value and, using parametric tests, the degrees of freedom (calculating variable and case numbers).

Although primarily about applied research, the following tips also work well in academia. Three issues are of central importance: The type of data, the target audience and the type of graphs being shown.

Data – absolute or averages?

Actually, you also need to consider this during your analysis. Even experienced analysts concern themselves about what is appropriate. Textbooks always make it look easy, but time after time, you will need to consider that apparently simple question, do I use the actual numbers or do I use a measure of central tendency (mean, median or mode)?

As each study is different, I doubt if there is a simple answer, but let us look at a few examples. If we wish to compare the industrial capacity of more than one country, wanting to know if one can produce more than another, then absolute numbers probably make more sense. On the other hand, if we are interested in the lives of individuals in those countries, for example, their earnings or ability to spend on

Data – absolute or averages?

items beyond subsistence level, then some type of average or indexing (collecting together 'baskets' of information and using a single scale), is probably more appropriate.

Another issue is the size. Let's start with small numbers. If you have been interviewing people or running focus groups, and there were only 10 people, would it really be appropriate to say "70% of respondents believed in the efficacy of this policy" when in fact those respondents were just a magnificent 7? I think I'd go for the absolute numbers.

Moving to larger numbers, think about government expenditure. Does the average person really know if 10 million dollars is a large or small amount by the standards of the day? Also, if comparing the expenditure with other countries, with different sizes of populations and different currencies, would not averages be more helpful? In many cases, one uses both sets of figures, but it helps to know which to emphasize in order for the audience to follow the logic of your study.

A few general points may help, but they are not eternal verities. For example, I may say "use the median" on certain occasions, but your employers may demand the mean at all times. Here are some general suggestions:

- Make it clear what type of data is being used.

- If using classification data (nominal/categorical/qualitative), you will typically show absolute numbers, although they may sometimes be accompanied by the proportions.

- If you use a measure of central tendency, do make it clear which you are using.

- Use the mode to represent the most common response.

- Use the median to represent 'lumpy' data, the sort with which we usually use non-parametric tests.

- Use the median where continuous data is skewed away from the normal distribution.

- Use the mean for the results from parametric tests.

It is fairly common practice to accompany your main figure with the standard deviation (SD), a standardized way of representing the dispersion from the mean, positively and negatively. This may be useful for experienced researchers, but I still think that the median is effective for skewed data, as it is insensitive to extremes in the data.

Different audiences

Try to consider the likely level of sophistication of your audience. Here I mean their statistical understanding, although their likely views on social policy may also be of relevance!

The reader or member of the audience could be the commissioners of the research, or a line manager, or members of the public. To some extent you will have to guess, but I suggest three rough levels of statistical understanding.

The sophisticated audience is likely to know at least as much about statistics as you do and probably a lot more. The intermediate audience may remember vaguely about significance levels (usually the critical value of $p < .05$). The unsophisticated audience will not understand the difference between the word significance in its statistical sense and its dictionary usage; p values will be meaningless.

Let's look at some of the different concepts and consider the likely audiences:

- The null hypothesis – in general, I would not use this outside of academia. 'Significant' and 'non-significant' will normally do.

- p values – '$p = 0.042$' is only for the most sophisticated audiences.

- Critical values – '$p < .05$' is ok for sophisticated audiences. Intermediate audiences may need a brief introduction on the occasion of its first usage, that it represents the probability of the effect being chance, lower figures suggesting that flukes are less likely and that being smaller than .05 indicates likely statistical significance.

Different audiences

- One-tailed and two-tailed hypotheses/tests – only sophisticated audiences will want to know about this and even for such an audience, a discussion of the expected direction would be helpful.
- Effect sizes – terms such as variance are only for sophisticated audiences, similarly r squared and other such statistics. 'Large', 'medium-sized' and 'small' effects would be suitable for an intermediate audience. Be even more sparing with unsophisticated audiences: mentions of particularly large and small effects will suffice, where they are of importance.
- Tests – sophisticated audiences will want to know which were used. Occasional usage may help intermediate audiences to believe that you know what you're doing, for example, "the result was significant to $p < .05$ (Mann-Whitney)". I would be inclined to omit this altogether when presenting to a lay audience.
- How you organized your data – in general, non-university audiences do not want to know about this, unless it is strictly of relevance to the study as a whole or you happen to think it important for a specific audience (maybe research commissioners, who paid for the work). You should always keep records, of course. In the case of your omitting outliers, a sophisticated audience should be informed, as they are likely to understand the relevance or otherwise of the data.

In general terms, the lay reader wants relevant results, ones with a bearing on the purposes of the research. It should not represent an academic thesis and while not an entertainment, should be readable and cogent.

Ah, says the worried, what happens if my audience is of *mixed levels of sophistication?* If you are fairly sure that your audience is highly variegated, and you believe that it is important that the lower level reader is not made to feel like a lemon (or whichever is your least favored fruit), then stratify your report. Perhaps put your most basic comment as the major part of a slide, followed by a more sophisticated comment in parentheses (e.g., "There was a significant difference between groups of self-employed respondents and those who were employed ($p < .05$ two–tailed).") If you know that you have even more fanatical stats-hounds in the audience, then consider footnotes.

Graphics

This is purely my own take on this. I try to limit my output on descriptive statistics to charts showing columns, charts showing horizontal bars and pie charts (leaving aside specialist charts such as those for correlations, factorial ANOVA and Kaplan-Meier plots for survival analysis). In general, I prefer columns for contrasting data, but bars come in useful if you have a lot of variables and/or lengthy titles, so that you need to spread down the page. I prefer to keep it simple, with one set of bars representing one effect, rather than layers or multiple meanings. Complex charts may mean something to you, because you've had your head in the data for extended periods; they may mean a lot less to your audience.

Pie charts are for exclusive data, where all the proportions are accounted for. If contrasting different groups, I generally prefer to have multiple pie charts, each representing a different group, rather than doughnuts and other concoctions.

Spreadsheets can also create simple correlation graphs (scatter plots), including a facility for adding a trendline. Other, quite complex graphs are also available to handle a variety of situations. If you choose to show one with layers, for example, be prepared to explain what it represents.

In general, I prefer to use graphics from a spreadsheet program rather than a specialist package. They are easier to tweak, adding titles, playing with the fonts and dealing with scales.

To a live audience!

Content needs to be limited – like short reports, only more so.

There are two particular things that will bore your audience and thus detract from your message: too much talking and information overload. Continuous talk is conducive to sleep or at best, lack of attention to what you think is important. The following practices are suggested for avoiding information overload and boredom.

- Try showing only one idea at a time, with a graph and the relevant statistics (depending on the audience) on one page.
- Try to keep slides interesting. Maybe have information sliding in from the side (not that I've ever mastered this). Avoid large slabs of text.
- Too much color can be distracting. Black and white is more effective than you might think.

While written reports require well-formed grammatical sentences, a presentation slide does not require all of the usual connecting phrases (although the voice-over does). The screen itself can have things like:

- "significant difference between the three social groups"
- "moderate effect size"
- "limitations in available data"
- "implications for research into housing conditions"

You then provide a commentary as you read from the screen. "We found a significant difference..." – including things of interest. "The lack of data pertaining to owner-occupied housing association complaints raises the question of how far this data can be generalized. Further research may be worthwhile in this area."

To a live audience!

Just repeating what you have written on the overhead chart is a terrible practice: your average member of the audience will wonder why they have turned up to listen. It is a good idea to prepare additional comments. At least do it in your head, but I think that rehearsing a couple of times is usually more effective.

For the purposes of smooth presentation and the avoidance of stage-fright, relatively inexperienced presenters may find it helpful to do a rehearsal in front of sympathetic colleagues. Be professional, addressing your audience at rehearsals as if they were your formal audience. This increases the chances of your being able to move into automatic mode when you are doing the real thing, unselfconsciously saying what you want to say.

You don't need to learn your words by heart. The screenshots are your prompts. Remember the few additional things you want to add; mentally associating them with the key phrases should help things to run smoothly. (And if it is your first time, don't dwell on the fact. Most of your audience will have done the same thing or will have to do so in the near future.)

In general, I would say that the key to successful reporting, oral or written, is effective categorization. Categorize, put the categories into a meaningful order and omit those which are likely to confuse or to bore unnecessarily. If some tedious things must be retained, then put them in a place where they are accessible but not center stage.

Chapter 11 – Factorial ANOVA and multiple comparisons

Factorial ANOVA deals with more than one effect, or 'factor', at a time. As with the one-way ANOVA, we can examine an effect and, with multiple comparison tests, its constituent conditions ('levels'). Where the one-way ANOVA considers only differences, factorial ANOVA also allows us to look at relationships between factors, called interactions. That we can examine both differences and relationships at the same time is because of the underlying statistical model, the General Linear Model (GLM). GLM underlies not only factorial ANOVA but also one-way ANOVA, t tests and regression.

Factorial ANOVA is generally used for two-way or three-way analyses. It can be applied to more than three factors at a time, but additional factors make it hard to interpret the results. *

---

*It is also possible, using the same logic in reverse, to apply ANOVA to only two conditions within a single variable, but the t test gives us sufficient information.

Typical case studies

Factorial analysis of variance – within subjects

A series of documentaries is shown to the same audience, measuring their attitudes toward immigration after each documentary. Each documentary is about the life of an individual immigrant, each time from a different ethnic background. The series was screened in two phases, two episodes at Christmas and two in April. The factors are the immigrants' backgrounds and the time of year.

Factorial analysis of variance – between subjects

Are schoolchildren in ability groupings more likely to perform well academically than those in mixed ability groupings. Does gender have a part to play?

Factorial analysis of variance – mixed design

The documentaries series is still seen by the whole audience, but we split the responses according to the ethnic background of the audience, which becomes a between-subjects factor.

Effect sizes for factorial ANOVA

Clark-Carter (1997) makes the following recommendations: 0.14 is large, 0.06 is medium and 0.01 is small. This is extrapolated into bandings by Kinnear and Gray (2004):

   Large: > .1   (more than 10% of the variance)
   Medium: 0.01 to 0.1 (1% to 10% of the variance)
   Small: < .01   (less than 1% of the variance)

Repeated Measures Two-Way ANOVA

We wish to review the effectiveness of a campaign against sexism in the workplace. HR staff have collected the data, which represents the number of recorded incidents in ten different offices. One of the factors is the intervention itself, divided into phases: before, during and after the campaign. The other factor is time, to see if differing moods in mornings and afternoons may affect responses.

Open the **TwoWayRepeatANOVA.csv** file.

Case	PreCampAM	PreCampPM	CampaignAM	CampaignPM	PostCampAM	PostCampPM
1	6	8	4	5	6	8
2	4	5	3	4	4	4
3	9	8	6	5	8	6
4	7	4	7	6	8	8
5	6	7	6	6	7	6
6	7	8	5	7	6	7
7	5	5	4	3	4	5
8	6	8	4	5	5	7
9	4	3	3	4	5	6
10	6	9	4	6	5	8

The data set looks similar to a One-Way Repeated Measures ANOVA, but each column in fact contains subdivisions of the data:

Case	Pre-campaign (1)		Campaign (2)		Post-Campaign (3)	
	AM (1)	PM (2)	AM (1)	PM (2)	AM (1)	PM (2)
1	6	8	4	5	6	8
2	4	5	3	4	4	4

The order is Factor 1 (1) with Factor 2 (1), then Factor 1 (1) with Factor 2 (1) and so on.

Press the ANOVA tab and select Classical / Repeated Measures ANOVA from the drop-down menu. At first, this is set up like the Repeated Measures one-way ANOVA.

Repeated Measures Two-Way ANOVA

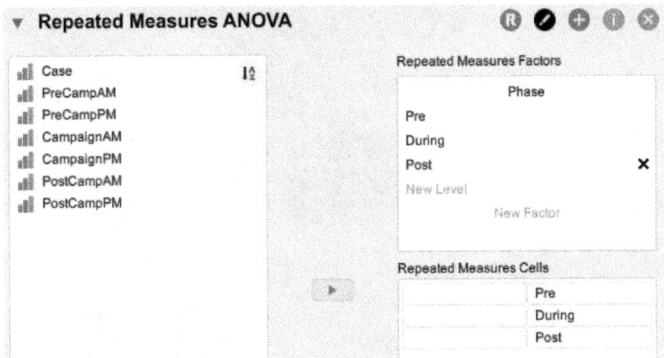

In the Repeated Measures Factors box, we replace RM Factor 1 with the name of the first factor (here, Phase) then adding in the conditions (Pre, During, Post) directly underneath the factor name.

Now, we do the same thing underneath for New Factor (using Time) and then its conditions/levels.

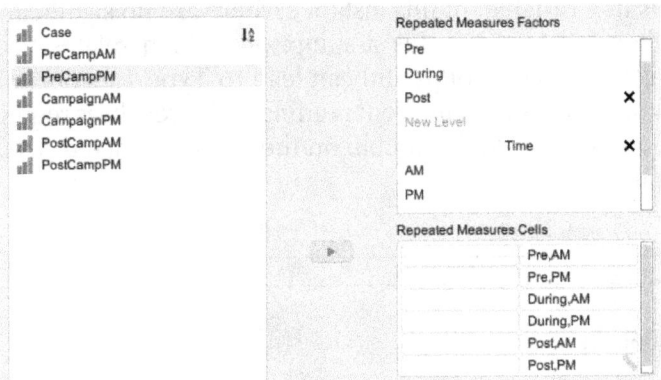

We can now see in the Repeated Measures Cells that permutations of the levels of the two factors have been automatically inserted on the right-hand side. The first level of the first factor has been conjoined with the first level of the second factor; then the first level of the first factor is matched with the second level of the second factor; and so on.

Repeated Measures Two-Way ANOVA

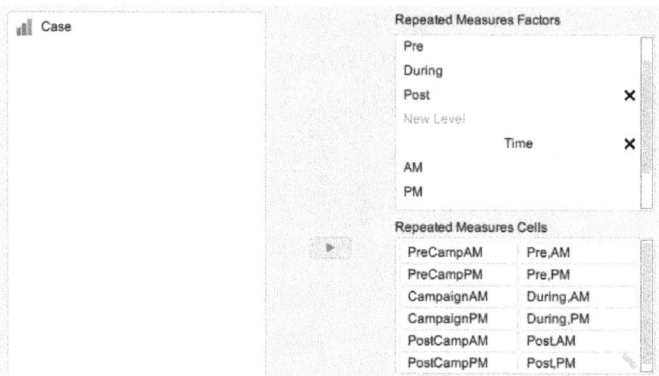

You then need to transfer the variables on the far left to the left-hand side of the Repeated Measures Cells. This example is straightforward, but as with other variants of factorial ANOVA, you need to be careful to ensure that each entry matches its logical pairing on the right.

Before analyzing the results, you should go to the Assumption Checks panel and select Sphericity tests. These apply when there are more than two levels in a variable; in this instance, there are no significant results. If your data fail to meet this assumption, which relates to variance pertaining to pairs of groups and can lead to Type 1 errors (mistakenly assuming significance), a different reading of the results is required. This is covered in the worked example on the two-way mixed ANOVA later in the chapter.

Within Subjects Effects

Cases	Sum of Squares	df	Mean Square	F	p
Phase	24.400	2	12.200	9.438	0.002
Residuals	23.267	18	1.293		
Time	4.817	1	4.817	1.883	0.203
Residuals	23.017	9	2.557		
Phase ✶ Time	0.133	2	0.067	0.159	0.854
Residuals	7.533	18	0.419		

The first factor, Phase, has a large F ratio and a p value of 0.002. The second factor is not significant – the time of day is clearly immaterial

Repeated Measures Two-Way ANOVA

– and the same can be said for the interaction between the two factors (Phase * Time). If you choose an effect size (I usually use Partial Eta Squared), you will see on the ANOVA table that the effect size for Phase is a very large .512, over 50% of the variance.

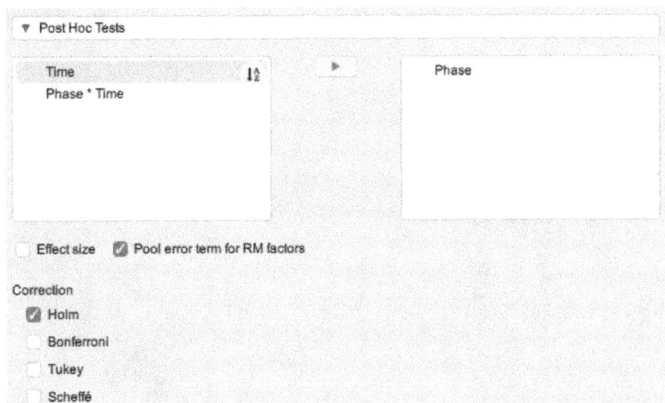

As the Phase main effect is significant, the Post Hoc Tests section has been opened. The relevant effect has been transferred to the right. I have used the Holm test; the tests and their potential usage will be discussed at the end of the chapter.

Post Hoc Comparisons - Phase

		Mean Difference	SE	t	p_{holm}
Pre	During	1.400	0.360	3.894	0.003
	Post	0.100	0.360	0.278	0.784
During	Post	-1.300	0.360	-3.616	0.004

Note. P-value adjusted for comparing a family of 3
Note. Results are averaged over the levels of: Time

While there is no relationship between the pre- and post- phases, there would appear to be significant differences between the campaign period and each of these phases. It's time for a chart.

Chapter 11 – Factorial ANOVA and multiple comparisons   181

Repeated Measures Two-Way ANOVA

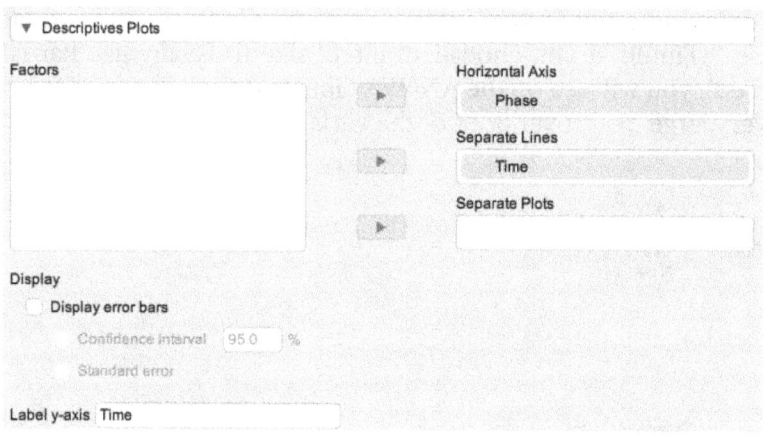

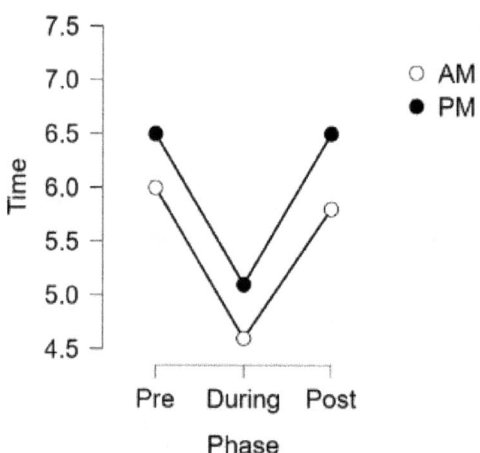

Three points arise from the chart. Firstly, we see that incidents are clearly lower during the campaign (but not only were they higher before, but they also rose to similar proportions afterwards). So the intervention would appear to have had an effect, but the stimulus did not have lasting effects. Secondly, the fact that both time lines (AM, PM) follow the same pattern indicates that the Phase factor is a 'global' effect. Thirdly, the lines do not intersect, not even tending towards it; there is no sign of interaction between the factors.

Repeated Measures Two-Way ANOVA

While we are on the subject of interactions, it is worth knowing that not only is it possible to get a significant interaction, but there are also times when only the interaction is seen to be significant. In one (real) educational survey, the main effects on native students' performance outcomes were supposed to be the levels of immigration at each school and the average parental education level of the immigrant students. Neither main effect seemed to influence performance significantly. On the other hand, there was an interaction between the two effects. Apparently, the parents of immigrant students were quite often graduates.

Generally, it is worth saying that all such plots should be examined with care. You may also want to select other panels for reporting and in-depth analysis.

Bayesian equivalent

Statistic		Quantification of evidence
Bayes Factor (BF10)	BF reciprocal (BF01)	
< 1	> 1	Noise
1 – 3	1 – 0.33	Weak
3 – 10	0.33 – 0.1	Moderate
10 – 20	0.1 – .05	Positive
20 – 150	.05 – .0067	Strong
> 150	< .0067	Very strong

This is a reminder of a suggested guide to Bayesian hypothesis reporting.

Using **TwoWayRepeatANOVA.csv**, select ANOVA / Bayesian / Repeated Measures ANOVA. The data entry is the same as in the classical version:

Repeated Measures Two-Way ANOVA

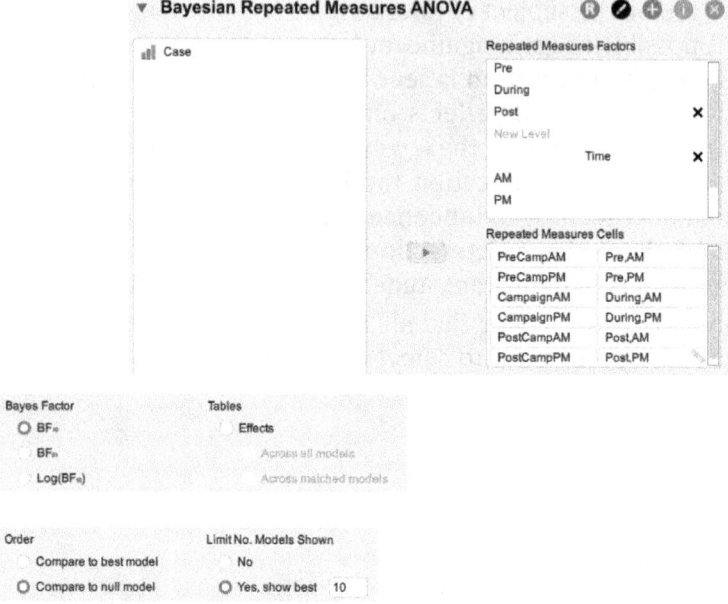

As previously, I have opted for 'Compare to null model' rather than the default 'best model' option.

Model Comparison

Models	P(M)	P(M|data)	BF_M	BF_{10}	error %
Null model (incl. subject and random slopes)	0.250	0.017	0.053	1.000	
Phase	0.250	0.512	3.143	29.353	1.711
Phase + Time	0.250	0.455	2.506	26.114	2.848
Time	0.250	0.016	0.048	0.903	5.370

Note. All models include subject, and random slopes for all repeated measures factors.

The output is rather oblique. As would be expected from the results of the frequentist test, Time has a Bayes factor of less than 1, any apparent effect likely to be noise, and Phase has a Bayes factor within the Strong credibility banding. The output does not show the interaction between Phase and Time, which would be shown as Phase * Time. What we do see is Phase + Time, which means the cumulative effect of the two variables together; the effect, however, has a somewhat smaller Bayes factor than Phase as a main effect, suggesting that the use of the two factors together is not a particularly parsimonious model.

Repeated Measures Three-Way ANOVA

Open the **ThreeWayRepeatANOVA.csv** file. Click the ANOVA tab and select Classical / Repeated Measures ANOVA. We are running a new program later in the year to deal with the same problem; maybe the lesson will sink in, or it's a different season, or maybe a new technique such as a video with role models may do the trick. The main points here are data input and interpretation of charts.

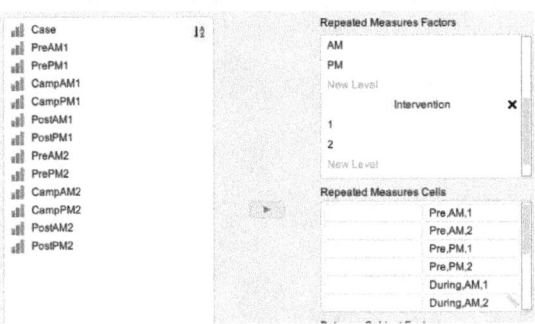

In terms of the source file, as you can see on the left-hand side, I have doubled the number of variables in the file, renaming the original ones with a '1' suffix, using the same names but with a '2' for the new set, representing the second intervention. Unlike the simpler two factor ANOVA, the variables do not match the right-hand permutations from top to bottom; *each variable needs to be put in carefully, one by one, to ensure correct matching.* Here you can see some of the matches:

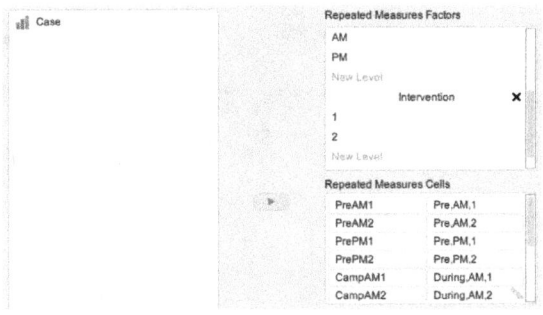

Repeated Measures Three-Way ANOVA

For Repeated Measures ANOVA, we can ignore the Between Subject Factors box, which is for the mixed design factorial ANOVA, to be discussed later in the chapter, where we have a hybrid of repeated measures (same subjects) and also between-subjects design (using groupings). Also ignore the Covariates box, which is for variables which we think are influential, but are not of interest to the study.

The output shows significant results for Phase again, but also for an interaction, Phase * Intervention, the latter being our new factor. You may want to check Effect Size. Charts are called for:

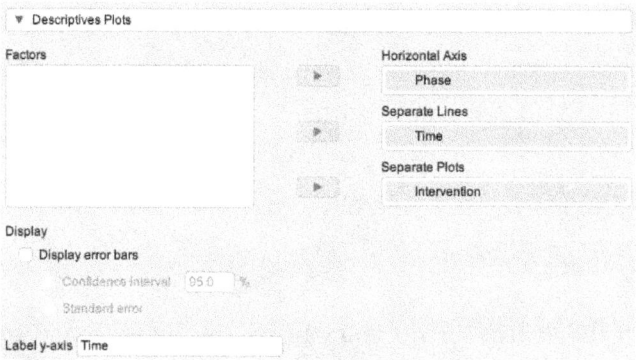

In order to examine the same effect as previously 'before and after', I have decided to have the same chart specifications as before, but with separate plots, intervention 1 and intervention 2.

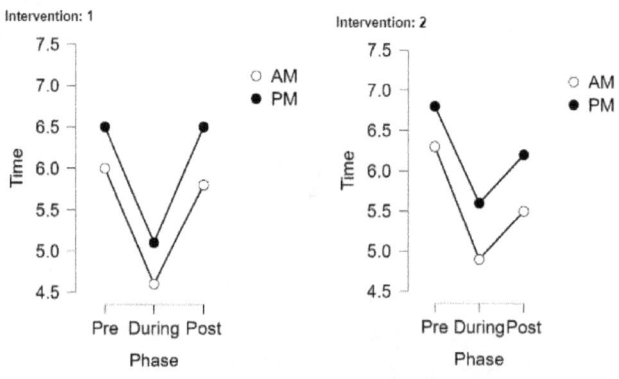

Repeated Measures Three-Way ANOVA

The charts suggest rather less of a bounce-back after the second intervention. On the other hand, there appear to be more incidents during the campaign than previously, and the post-campaign reduction is not as drastic as may have been hoped.

Let's see a chart for the interaction:

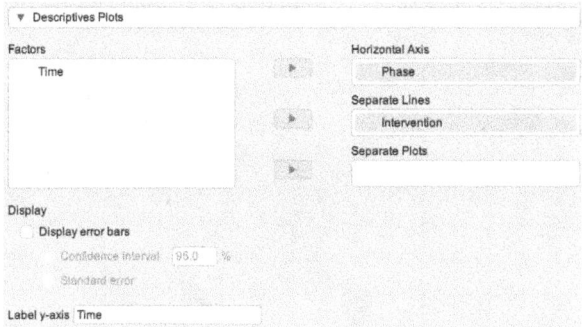

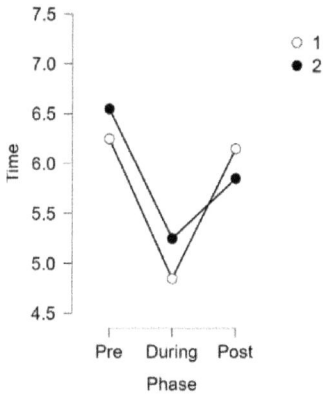

Note the intersection of the lines, indicating an interaction of the two effects. If we had a movement towards intersection rather than actual crossing, this would suggest a trend towards interaction. If you add Time to the interaction, the same effect will be apparent regardless of the different times.

Repeated Measures Three-Way ANOVA

Bayesian equivalent

Using **ThreeWayRepeatANOVA.csv**, select ANOVA / Bayesian / Repeated Measures ANOVA.

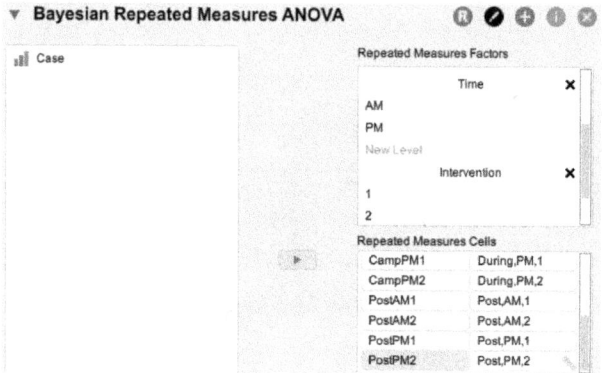

Ensure that Bayes Factor option is set to the default BF10. The Order option should be set to 'Compare to null model'. The computation may take some time... *

Models	P(M)	P(M|data)	BF_M	BF_{10}	error %
Null model (incl. subject and random slopes)	0.053	0.016	0.292	1.000	
Phase + Intervention + Phase ✻ Intervention	0.053	0.205	4.632	12.824	5.948
Phase + Time + Intervention + Phase ✻ Intervention	0.053	0.203	4.579	12.707	12.280
Phase + Time	0.053	0.126	2.585	7.868	16.339
Phase	0.053	0.107	2.163	6.723	4.905
Phase + Time + Intervention + Phase ✻ Intervention + Time ✻ Intervention	0.053	0.062	1.181	3.858	8.404
Phase + Intervention	0.053	0.061	1.160	3.793	8.527
Phase + Time + Phase ✻ Time	0.053	0.052	0.997	3.289	66.467
Phase + Time + Intervention	0.053	0.046	0.861	2.861	9.083
Phase + Time + Intervention + Phase ✻ Time + Phase ✻ Intervention	0.053	0.032	0.595	2.006	40.497

Note. All models include subject, and random slopes for all repeated measures factors.
Note. Showing the best 10 out of 19 models.

The Phase ✻ Intervention interaction duly appears, together with the Phase and Intervention variables, within the Positive banding. The

---

*You may find that your computer takes some time to calculate this; as mentioned earlier, processing speeds only allowed Bayesian statistics to be used relatively recently.

Repeated Measures Three-Way ANOVA

fairly minimal difference in Bayes factors between this and the next line, including Time, suggests that the second model is not parsimonious. A similar consideration can be made about the next two models, Phase + Time and Phase; Time only seems to make a limited contribution.

Analysis of Effects

Effects	P(incl)	P(excl)	P(incl|data)	P(excl|data)	BF$_{incl}$
Phase	0.737	0.263	0.953	0.047	7.194
Time	0.737	0.263	0.603	0.397	0.543
Intervention	0.737	0.263	0.684	0.316	0.772
Phase * Time	0.316	0.684	0.127	0.873	0.316
Phase * Intervention	0.316	0.684	0.531	0.469	2.457
Time * Intervention	0.316	0.684	0.117	0.883	0.288
Phase * Time * Intervention	0.053	0.947	0.003	0.997	0.046

You may find the Effects option helpful.

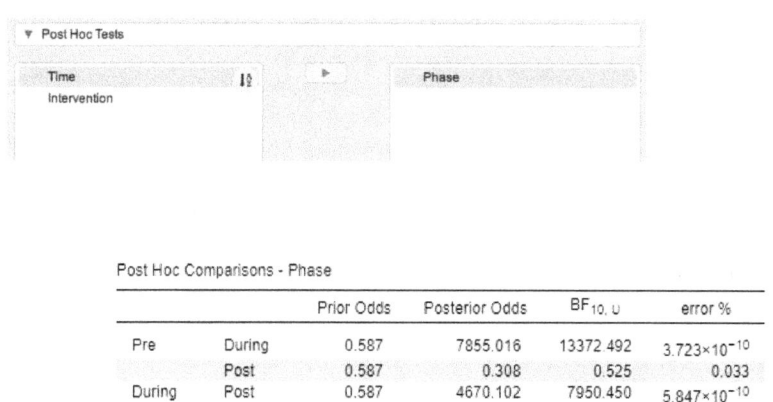

Post Hoc Comparisons - Phase

		Prior Odds	Posterior Odds	BF$_{10, U}$	error %
Pre	During	0.587	7855.016	13372.492	3.723×10^{-10}
	Post	0.587	0.308	0.525	0.033
During	Post	0.587	4670.102	7950.450	5.847×10^{-10}

This provides a similar result to the frequentist post hoc test.

Between-Subjects ANOVA

Case	Broadcast	Politics	Gender	Rating
18 18	A	Republican	female	7
19 19	A	Republican	female	6
20 20	A	Republican	female	8
21 21	B	Democrat	male	3
22 22	B	Democrat	male	4
23 23	B	Democrat	male	6
24 24	B	Democrat	male	1

Open the **BetweenANOVA.csv** file. Here is a snapshot from the data. Each person has been shown one of two closed circuit television broadcasts about a current political issue (A or B) or a Control documentary with no current implications. Political party and gender have also been recorded, along with a measure of their intended activism on the issue.

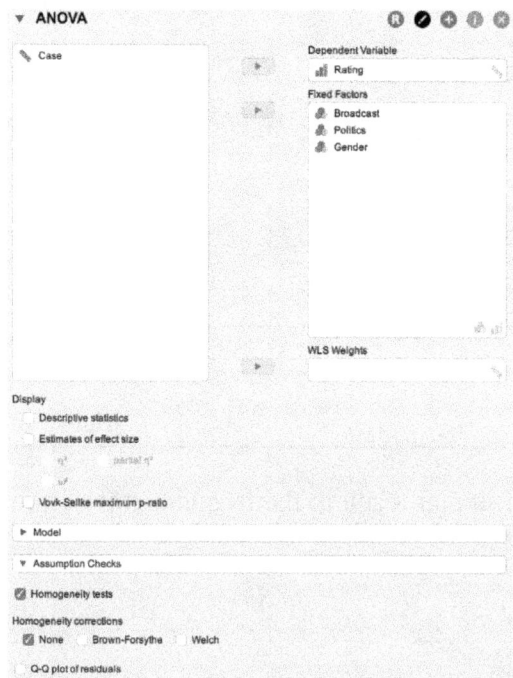

Press the ANOVA tab, selecting Classical / ANOVA, and here we have a three-way ANOVA.

ANOVA - Rating

Cases	Sum of Squares	df	Mean Square	F	p
Broadcast	47.433	2	23.717	6.858	0.002
Politics	25.350	1	25.350	7.330	0.009
Broadcast ∗ Politics	4.300	2	2.150	0.622	0.541
Gender	6.017	1	6.017	1.740	0.193
Broadcast ∗ Gender	0.233	2	0.117	0.034	0.967
Politics ∗ Gender	0.817	1	0.817	0.236	0.629
Broadcast ∗ Politics ∗ Gender	0.433	2	0.217	0.063	0.939
Residuals	166.000	48	3.458		

Here, only the Broadcast and Politics variables are deemed significant.

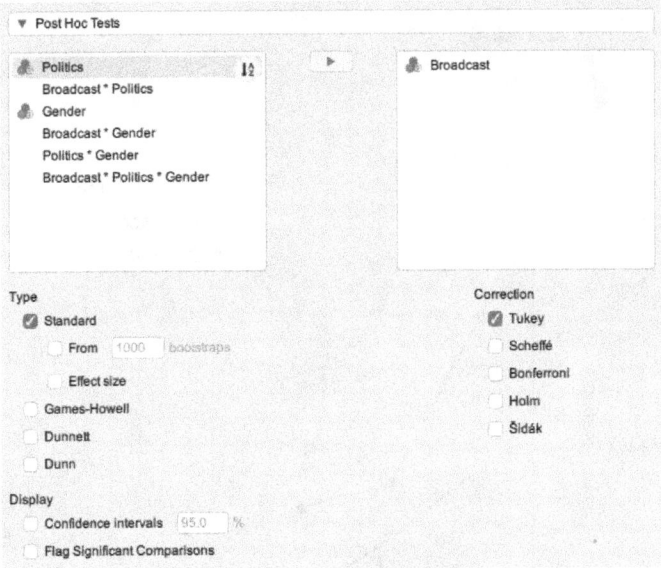

We only use a comparison test with Broadcast as it has three conditions, so pairings make sense. I am interested in the effect of Broadcasts on intentions to engage and am particularly interested in how they are considered by people with different political allegiances. The Tukey test is the default here.

Between-Subjects ANOVA

Post Hoc Comparisons - Broadcast

		Mean Difference	SE	t	p_tukey
A	B	1.550	0.588	2.636	0.030
	Control	2.100	0.588	3.571	0.002
B	Control	0.550	0.588	0.935	0.621

Note. P-value adjusted for comparing a family of 3
Note. Results are averaged over the levels of: Politics, Gender

Broadcast A differs significantly from the other two conditions.

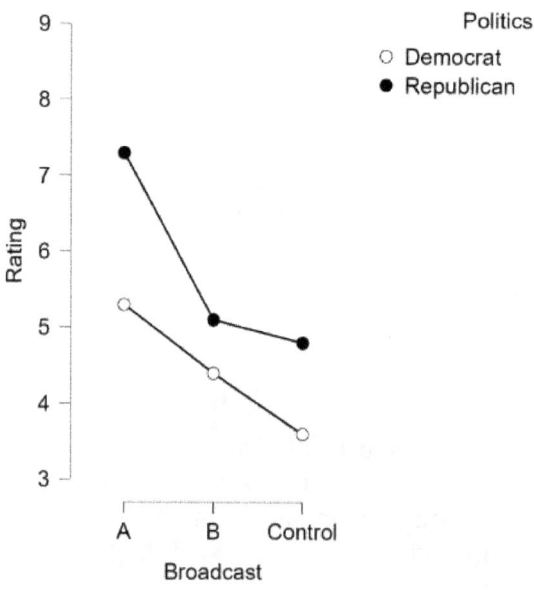

The chart shows quite a lot. Firstly, Republicans seem much more likely to be activists over these issues than Democrats, and this seems independent of the broadcast types. Looking at the leftmost node of the Democrat ratings, we can also see that Broadcast A elicits superior ratings regardless of political leanings, as it rates higher than all B and Control nodes; this is supported by the comparison tests, with Broadcast A significantly different from B and from the Control showing. (You may also enjoy trying out the Raincloud Plots option.)

Bayesian equivalent

Open **BetweenANOVA.csv**, select ANOVA / Bayesian / ANOVA.

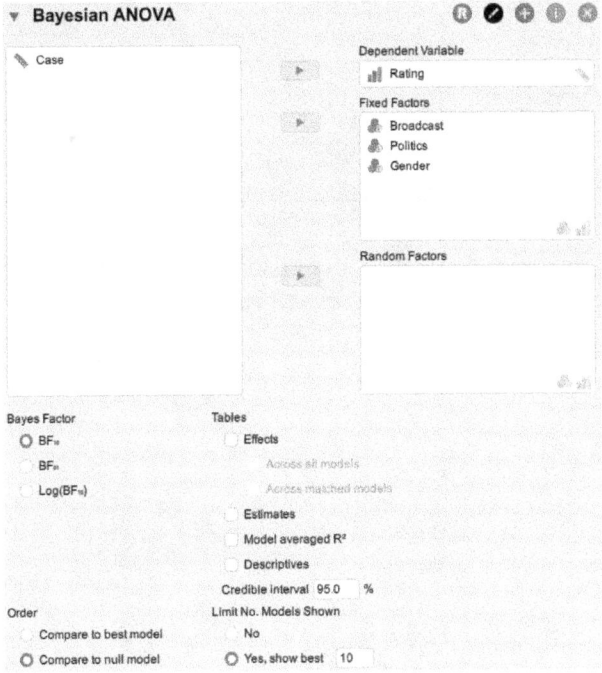

The data is entered in exactly the same way as above, but remember to 'Compare to best model'.

Between-Subjects ANOVA

Model Comparison

Models	P(M)	P(M|data)	BF$_M$	BF$_{10}$	error %
Null model	0.053	0.003	0.061	1.000	
Broadcast + Politics	0.053	0.335	9.051	99.416	1.167
Broadcast + Politics + Gender	0.053	0.199	4.460	59.002	1.028
Broadcast + Politics + Broadcast ✻ Politics	0.053	0.110	2.230	32.758	2.292
Broadcast + Politics + Gender + Politics ✻ Gender	0.053	0.072	1.389	21.292	2.229
Broadcast + Politics + Gender + Broadcast ✻ Politics	0.053	0.063	1.210	18.720	3.091
Broadcast	0.053	0.052	0.996	15.572	0.008
Broadcast + Politics + Gender + Broadcast ✻ Gender	0.053	0.043	0.806	12.738	3.619
Broadcast + Gender	0.053	0.029	0.530	8.494	2.018
Broadcast + Politics + Gender + Broadcast ✻ Politics + Politics ✻ Gender	0.053	0.026	0.472	7.586	9.160

Note. Showing the best 10 out of 19 models.

The output is rather gnomic. The Effects table is more enlightening:

Analysis of Effects - Rating

Effects	P(incl)	P(excl)	P(incl|data)	P(excl|data)	BF$_{incl}$
Broadcast	0.737	0.263	0.973	0.027	13.008
Politics	0.737	0.263	0.908	0.092	3.515
Broadcast ✻ Politics	0.316	0.684	0.223	0.777	0.620
Gender	0.737	0.263	0.487	0.513	0.339
Broadcast ✻ Gender	0.316	0.684	0.089	0.911	0.211
Politics ✻ Gender	0.316	0.684	0.122	0.878	0.302
Broadcast ✻ Politics ✻ Gender	0.053	0.947	0.002	0.998	0.035

This produces a clearly similar response to the frequentist results. Only Broadcast and Politics have significant results.

Mixed ANOVA

	Case	School	Mathematics	English	Science
1	1	state	80	82	78
2	2	state	65	67	64
3	3	state	50	58	45
4	4	state	68	69	70
5	5	state	63	66	63
6	6	state	57	56	58
7	7	private	84	83	84
8	8	private	70	75	71
9	9	private	70	76	72
10	10	private	57	62	58
11	11	private	46	60	42
12	12	private	55	64	51

Open the **MixedANOVA.csv** file. There is a grouping according to type of school, but all schools take examinations in the same subjects ('repeated measures'): Mathematics, English and Science.

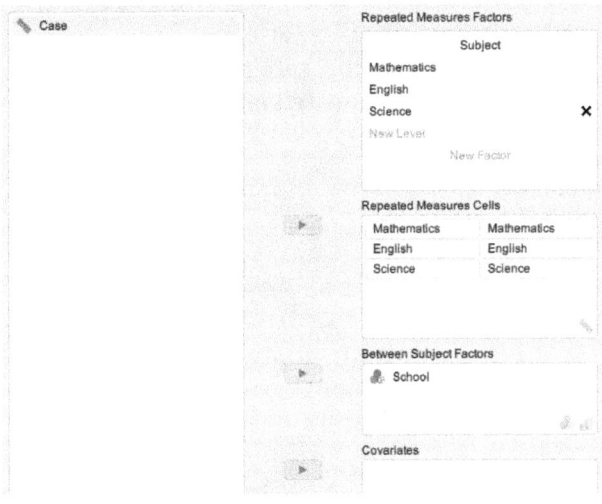

Mixed ANOVA

Click the ANOVA tab and select Classical / Repeated Measures ANOVA from the drop-down menu. Here we have three subjects entered as Repeated Measures Factors and one grouping in Between Subject Factors. This is a two-way mixed ANOVA, as there is one Repeated Measures (RM) factor, the academic subject, and one Between Subject factor, the type of school.

If you wanted to extend the design, there are two ways of creating a three-way ANOVA: you could put another grouping factor in the Between Subject Factors box, gender perhaps, or you could add another factor to Repeated Measures Factors, maybe entering the results according to both final year and test results from a previous year. It is of course possible to have more than three factors, but the more of them you have, the harder it is to interpret the results.

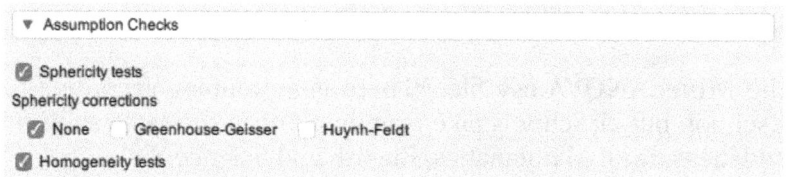

Because we have a between-subjects factor, we need the homogeneity test, checking that there are similar variances across the conditions. We also want the sphericity test; the sphericity assumption is rather like homogeneity of variance for repeated measures.

Test for Equality of Variances (Levene's)

	F	df1	df2	p
Mathematics	1.075	1	10	0.324
English	0.355	1	10	0.564
Science	1.486	1	10	0.251

Mixed ANOVA

Test of Sphericity

	Mauchly's W	Approx. X²	df	p-value	Greenhouse-Geisser ε	Huynh-Feldt ε
Subject	0.161	16.460	2	< .001	0.544	0.559

The sphericity test, Mauchly's test, is significant. We need to take a different reading of the results, providing a slightly more conservative p value to avoid Type 1 errors. Sphericity corrections are available in the Assumption Checks section. Some researchers use the Greenhouse-Geisser correction on all such occasions, others use the Huynh-Feldt.

My advice is to return to the readings shown to the right of the Mauchly result; if the Epsilon statistic (ϵ) for Huynh-Feldt is greater than .75, use Huynh-Feldt; otherwise, use Greenhouse-Geisser (Girden 1992). So in this instance, we use the latter.

Sphericity corrections
☐ None ☑ Greenhouse-Geisser ☐ Huynh-Feldt

Within Subjects Effects

Cases	Sphericity Correction	Sum of Squares	df	Mean Square	F	p
Subject	Greenhouse-Geisser	187.056	1.087	172.036	8.988	0.011
Subject * School	Greenhouse-Geisser	28.167	1.087	25.905	1.353	0.274
Residuals	Greenhouse-Geisser	208.111	10.873	19.140		

Between Subjects Effects

Cases	Sum of Squares	df	Mean Square	F	p
School	12.250	1	12.250	0.031	0.863
Residuals	3915.389	10	391.539		

The reading, as corrected, is shown, with an adjusted p value of 0.011 for the Subject main effect. Neither the school main effect nor the interaction are deemed to be significant.

As the Subject effect is of interest, we can go to the Post Hoc Tests section to look at the pairings.

Mixed ANOVA

Post Hoc Tests

Post Hoc Comparisons - Subject

		Mean Difference	SE	t	p_{bonf}
Mathematics	English	-4.417	1.317	-3.354	0.009
	Science	0.750	1.317	0.570	1.000
English	Science	5.167	1.317	3.923	0.003

Note. P-value adjusted for comparing a family of 3
Note. Results are averaged over the levels of: School

I have chosen to view the Bonferroni correction, rather than the Holm, for a reason. Given that our data is in breach of an assumption for repeated measures tests, it may be wiser to choose a more conservative test. We shall cover the subject of multiple comparisons shortly.

Mixed ANOVA

Bayesian equivalent

Open **MixedANOVA.csv**.

The Mathematics variable has been allocated to the Ordinal category (with three bars). This needs converting to Scale in order to run on this test; you do this by clicking on the category symbol (the bars), and opting for Scale (denoted by a rule symbol).

Select ANOVA / Bayesian / Repeated Measures ANOVA.

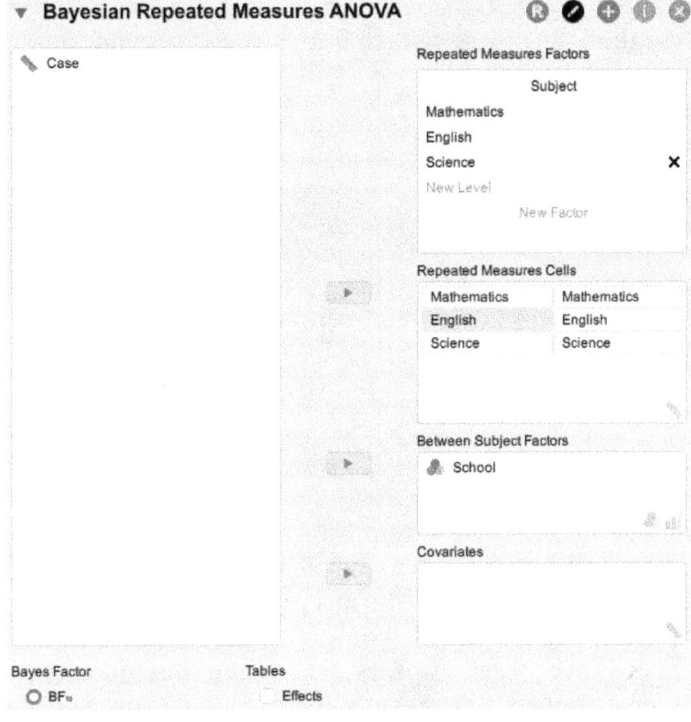

Mixed ANOVA

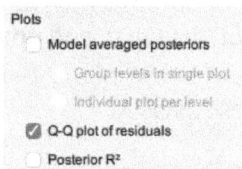

Model Comparison

Models	P(M)	P(M|data)	BF_M	BF_{10}	error %
Null model (incl. subject and random slopes)	0.200	0.022	0.092	1.000	
Subject	0.200	0.439	3.125	19.522	0.712
Subject + School	0.200	0.330	1.972	14.698	2.363
Subject + School + Subject ∗ School	0.200	0.193	0.956	8.585	2.883
School	0.200	0.016	0.065	0.707	1.739

Note. All models include subject, and random slopes for all repeated measures factors.

We can use the Q-Q plot as part of our testing of assumptions.

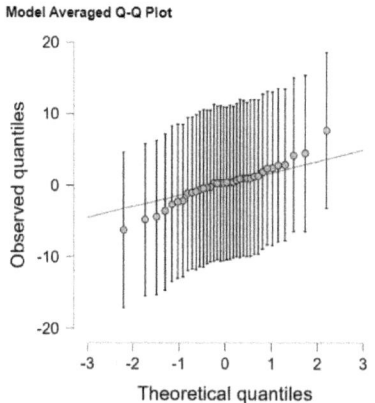

The points deviate to some extent from the reference line. It is at least a warning to avoid pronouncements about the alternative hypothesis being 16 times more likely than the null hypothesis, or even being within the Positive banding (10–20). It would be better to indicate that the Subject alternative hypothesis appears to lie within the Moderate banding (3–10) at least, reporting the results and explaining your decision.

Multiple comparisons

Also known as post hoc tests, corrections or adjustments, these tests are designed to prevent Type 1 errors, the assumption of a significant result when none exists. The concern is that if you have several pairings, there is the chance that one or more results will only appear to be significant through a fluke.

The multiple comparison tests may be used in a planned or unplanned way. With planned comparisons, you start off with a well-considered analysis of variance, including the expectation that particular pairings will require subsequent analysis. With post hoc (after the event) comparisons, you follow up a large but significant ANOVA result with wall-to-wall coverage of all possible pairings. I have provided two rather extreme models, as everyday practice usually falls between these, but the principle is worth considering for the purposes of this discussion.

Before choosing between comparison tests, liberal and conservative, I would like to introduce the idea that we should use none. The position that the ANOVA is an 'omnibus' test from which only the overall result could be accepted was defended by the great statistician R. A. Fisher (1935). Criticism of comparison tests persisted for quite some time (Nelder, 1971; Plackett, 1971; Preece, 1982).

Prior to the days of widespread computing, we did all the tests by hand, which was time-consuming but kind of fun (depending on one's view of life). Textbooks often didn't include multiple comparisons (for example, Greene and D'Oliveira, 1982). Adding hugely to the amount of time expended on calculations, it is not too surprising that multiple test usage was not widespread before computer usage (Parker, 1979), but well into the day of the personal computer, there was still the view that "multiple comparison methods have no place at all in the interpretation of data" (Nelder, 1999).

Perhaps the widespread incorporation of comparison tests into computerized statistical packages has led to a general acceptance of their use, but here is one group of researchers fulminating against the 'wickedness' of the use of a liberal test: "the true interpretation of the data was

Multiple comparisons

submerged in the swamp of significance statements" (Mead *et al*, 2012). The criticism of unreasoned usage still stands.

There is now little opposition to the use of planned tests. Where there are just a few anticipated pairings, it is generally considered reasonable not even to use the correction tests to be discussed in this chapter. As a 'no correction' option, *t* tests could be used. If you are using the Friedman test, you can use Wilcoxon tests of each pairing; after a Kruskal-Wallis test, use Mann-Whitney tests.

Unplanned tests (*post hoc*) are still very controversial (Games, 1971; Sato, 1996). Pearce (1993) considers them overused, because of automated computerized usage (not in JASP) and a lack of desire to specify contrasts. While unplanned tests are said to be less powerful than planned tests (Day and Quinn, 1989), I would suggest an intermediate view: unplanned tests could be used to generate new hypotheses, even though they should still be viewed with considerable skepticism. A more radical view is that of Hess and Olejnik (1997), that ANOVA should be abandoned in favor of focused hypothesis testing.

On the assumption that we wish to carry on using ANOVA and multiple comparison tests, we have another little controversy to consider. In this book, I have only followed up significant ANOVA results with comparison tests. Many consider follow-up from non-significant ANOVA tests to be a form of 'dredging' (for example, Wyseure, 2003). This view is challenged as a misconception by other researchers (for example Huck, 2008), who believe that comparison tests may be used regardless of whether or not the overall effect was significant.

Having considered these matters, let us consider the tests themselves. Some of the time, the different tests will give similar results. At other times, they differ. The following descriptions are only general; because the tests use different techniques, their apparent 'liberalism' or 'conservatism' should not be viewed as forever accurate.

In general, the **Holm** is considered the most liberal of the tests provided (also known as 'more powerful'); it is most likely to provide a lower p value, and thus more likely to pick up 'significant' results. This also means that it is slightly more likely to commit a Type 1 error, wrongly attributing significance to an effect. The Holm can be used for both repeated measures and between-subjects ANOVA.

The **Scheffé** is generally the most conservative of the tests. It will avoid Type 1 errors easily, but can occasionally be prone to Type 2 errors, wrongly considering results to be non-significant. Even if you have chosen this test, it is probably also worth checking the results of a Bonferroni test at the same time; every now and again, the Scheffé can be completely adrift from other test results. Results quite often differ, but they should not differ wildly. The Scheffé should only be used for between-subjects ANOVA.

The **Bonferroni** is a traditional conservative test. While not as tough-minded as the Scheffé, it is considered to be a little too strict for many occasions (Rice, 1989), although is still used in several textbooks (for example, Kinnear and Gray, 2004). The Bonferroni can be used for both repeated measures and between-subjects ANOVA.

Tukey is a reasonable intermediate test, not as strict as the Bonferroni, but not as liberal as the Holm. Traditionally used and the default in JASP for between-subjects ANOVA, the Tukey is the most widely used test (Tsoumakas *et al*, 2005). The Tukey should only be used for between-subjects effects. You may have seen it in the Repeated Measures ANOVA dialog; it can be used in a mixed ANOVA, referring to a variable in the Between Subject Factors box.

Dallal (2012) suggests the use of non-adjusted tests such as t tests for planned tests. Scheffé could be used where you want to use all of the comparisons you can think of, but Dallal is scathing about possibly missing an effect for the sake of comprehensiveness. Tukey is the comparison test he most often uses.

Hilton and Armstrong (2006) stress the importance of the purpose of the investigation in deciding upon test usage. "If the purpose is to decide which of a group of treatments is likely to have an effect, then it is better to use a more liberal test.. in this scenario it is better not to miss an effect. By contrast, if the objective is to be as certain as possible that a particular treatment has an effect then a more conservative test..

would be appropriate." Hilton and Armstrong consider Tukey's test to fall between the extremes.

They also believe (as does Dallal 2012) that "none of these methods is an effective substitute for an experiment designed specifically to make planned comparisons between the treatment means". You have probably heard this before, but it is always worth keeping in mind, that good research design makes analysis a lot easier and more efficient.

Other tests

In JASP's (between-subjects) ANOVA, you will find four more of these tests. The **Šidák** is probably here 'on demand', as it appears in various commercial software packages. It is only slightly more liberal than the Bonferroni correction, but it assumes that each comparison is independent of the others.

The other three are specialist tests. **Games-Howell** is used when the homogeneity assumption has been violated. The **Dunnett** test is a powerful test which is used to compare all of the groups with one group; this generally means that various treatments are compared with a control. The **Dunn** is a non-parametric test which may be used after a Kruskal-Wallis test; it is only suitable for a small number of pairings. The **Conover** test accompanies the Friedman test.

Thinking point

Are you using a test, or any other form of technology, just because you can? Apart from advantages and disadvantages, irrelevances can obscure more important issues.

Chapter 12 – A taste of further statistical methods

This book is for beginners, but nevertheless a lot of useful research can be conducted with the methods previously discussed. I offer here, however, a taste of what more advanced methods can do. In order not to create confusion, only a few types of test will be considered, generally building upon what you have already learned. Most of these are available in JASP's default menus or in the modules menu (use the plus symbol at the top right).

Data reduction methods

Principal components analysis and factor analysis

PCA and **Factor analysis** both take a relatively large number of variables, generally represented in your raw data as columns, and reduce them to hopefully comprehensible groups of variables.

In PCA, these groups, known as *components*, are created without any theory or rationale. You may for example be developing a questionnaire and find that the pilot project (always run a pilot project) reveals that respondents really don't want to answer so many questions. Creating components would allow you to eliminate some of the questions, or to consolidate into core questions.

Data reduction methods

In FA, the groups are called *factors*. In **exploratory factor analysis** (EFA) you are essentially reducing responses to find out how many key ideas emerge from the data, what they mean, and how the ideas interrelate. In **confirmatory factor analysis** (CFA), a little more advanced, you already have a good idea of the theory and you want to see how well measured variables represent the known constructs.

PCA is a non-parametric test and thus has less statistical assumptions to be met. FA has rather a lot of assumptions. While some researchers are adamant that PCA and EFA serve distinct purposes, others advocate the use of PCA for both empirical and theoretical usage. For example, Stevens (2009) claims that both techniques often offer similar results. While PCA may not be theoretically based, it can be used to similar effect (both offer 'rotation' these days, a story for another time or place). There are difficulties in using either technique, but a "a good PCA or FA makes sense. A bad one does not" (Tabachnick and Fidell, 2007). If you are reducing the number of variables, fine. If you are exploring a theory, also fine. But,

> "exploratory FA is not, or should not be, a blind process in which all manner of variables or items are thrown into a factor-analytic 'grinder' in the expectation that something meaningful will emerge. .. GIGO ... garbage in, garbage out." Pedazhur and Schmelkin (1991).

Cluster Analysis

This also pertains to data reduction, but its focus is upon the *rows* of a matrix, representing cases. You may want to see if attitudes to an issue are different, say, between people of different ages or from different parts of the country. It is not unknown for groups to become newly constructed from surveys, as different types of consumer, elector or social or personality type. These are often referred to as 'objects'. So in an inversion of factor analysis, we look not at core constructs but at the behaviour of clusters of individuals, or groups of data (Everitt *et al*,

2011). Hierarchical clustering is usually the technique with which to start.

Cluster analysis is best suited to empirical work. We usually don't have a governing theory, but think about what the clusters mean when we see them. We can follow up initial exploration with different ways of viewing the clusters, but it is really a descriptive method, not an inferential one. Statistically speaking, it is not proof of anything!

While statistically of little use, cluster analysis is an intensely mathematical tool. It is what is called an 'unsupervised' classification method: this means that you let the machine get on with finding any hidden structure. By contrast, a 'supervised' method is basically where you have created a 'training' model against which to try out fresh data.

For an objective assessment of differences between groups, something like ANOVA or logistic regression would make sense.

Logistic regression

At times, you are going to find that much of your research data consists of zeros and ones (yes / no) and other categorical data. Although categorical (or qualitative) methods have their place, when you have relatively rich data sets, you are likely to want to use methods which can sensitively weigh up the relative contributions of different variables. Logistic regression is popular among researchers because of its versatility, incorporating categorical, ordinal, or continuous data, often in combination.

Methodologically, this is different from other forms of regression that you have used. The logic of most of the previous tests is reversed. In ANOVA, t tests and correlations, a condition or grouping is the independent variable, with the dependent variable as a continuous or ordinal measure. Here, the dependent variable is categorical, sometimes known as the 'output' or 'classifier'. Put another way, we are seeing if data fits into certain discrete categories.

This technique is also referred to as logit regression or the logit model. It should be noted that interpretation of the output will be rather unfamiliar at first.

Survival analysis

Although survival analysis takes its name from medical research (people are quite interested in mortality), it has many other names dependent on context. It started with seventeenth century studies of risk, and patterns of longevity and mortality; insurance and annuities came into being in an organized way, using life tables as the foundations of actuarial work. Within engineering, 'reliability analysis' or 'reliability theory' studied how long it took for weapons to fail. In sociology and economics, you may find it referred to as 'event history analysis', 'event structure analysis', 'duration analysis' and 'duration modeling'.

I use the term 'survival analysis' here because it is the most commonly used term in the statistical literature; you would refer to this if you were to study the subject further. **The time to events** is a rather better term, given its explanatory clarity. We are looking at how long it takes for an **event** to happen. Statistically and practically speaking, that does not have to be a negative event such as death. It could be the point in time after diagnosis that a patient is wheeled into the operating theatre. An event could be positive or negative: puberty, deaths, graduation, dropping out, marriages, divorces, new jobs and lost jobs, the start of strikes or perhaps their cessation. The techniques are incredibly adaptable: you could study car crashes, violent incidents and perhaps the successful fruition of business or management operations. Perrigot *et al* (2004) examine the history of failures within business franchises in order to study what works organizationally and what does not.

Statistically, we record the number of days, or weeks, or years from a baseline time to an event. **Life tables** are typically used over years, handling data at intervals.

Over days, weeks, or months, the **Kaplan-Meier** survival function (also known as the Kaplan-Meier curve) provides a chart of considerable value. The curve's steepness shows the likelihood of an event, whether or not the effect accelerates or decelerates over time, and even the differences between groups of cases under different conditions. The Kaplan-Meier even takes into account cases of loss to follow-up (also known as censoring). Unlike life tables, Kaplan-Meier calculates each time an event takes place.

There are more advanced techniques within this approach, including the *Cox model* (also known as the 'Cox proportional hazards model') which takes into account multiple factors, and the Weibull and exponential distributions. In general the inexperienced researcher should stick to Kaplan-Meier and life tables, which are relatively free from assumptions.

Reliability

If you are creating a psychometric test or a questionnaire, it is not enough to be **valid**, that is, proven to do what it is supposed to do. If it is not **reliable**, that is consistent, then validity is undermined. To be reliable, a measure needs to work again and again. This is why we have pilot projects: they allow us to check that our metrics are reliable before we go ahead and use them in a full study or survey.

Internal reliability

This is also referred to as **internal consistency**: the contents of the related variables need to be consistent with each other. This internal consistency is traditionally measured by Cronbach's α (alpha), although these days McDonald's ω (omega) is favoured (Hayes and Coutts, 2020; Goodboy and Martin, 2020). JASP's Reliability module includes what it calls Unidimensional Reliability, which uses McDonald's ω by default. The option called Individual Item Statistics is of particular value for deciding on items to be removed.

Test-retest reliability

Are measurements consistent over time?

Inter-rater reliability

Are measurements consistent across different assessors?

Meta-analysis

The analysis of the pooled results from available research studies, in order to come to a conclusion about the area of interest. Among other qualifications, it should be noticed that the accuracy of meta-analysis tends to depend on the similarity of the evidence; not unusual within research, comparing like with like is important for coherent results.

ANCOVA – some words of warning

ANCOVA, the **analysis of covariance** is quite similar in function to ANOVA (and is a between-subjects design, not repeated measures). Covariance is how the variables change together. ANCOVA incorporates *covariates*, variables that are influential but not of interest. A combination of ANOVA and regression, it 'controls for' unwanted variables. Typically, the test is used to iron out flaws within quasi-experiments, removing statistical 'noise' for more accurate results than those provided by ANOVA. In practice, you get a read-out of values similar to those from ANOVA, but with values for the factors and the covariates; the most important point is that the values for the factors will have changed.

Achieving more accurate estimates of a factor's influence on the variance, ANCOVA is also used to try to remove the effects of fixed groups from a study (for example, administrative, managerial and manual staff) in order to study the overall effect. But before you rush off and use this,

ANCOVA – some words of warning

I would first of all like to introduce some methodological issues which might make you think again.

Assumptions for data

There are the usual assumptions for using parametric tests, those of continuous data, normal distribution and, for different numbers in different conditions, homogeneity of variance. An additional assumption is the need for the covariate to have a linear relationship with the dependent variable; this can be measured using a scatter plot.

Another assumption, arguably the most important of the additional assumptions, is that of homogeneity of regression: the dependent variable and the covariate must not be over-correlated (yes, this test wants to have its cake and eat it; Goldilocks' approach to porridge comes to mind).

Regression lines for the covariate across the different groups need to be parallel, neither crossing each other nor getting too close to each other. The covariate should be unrelated to the dependent variable; this should be checked during the design stage, not as a result of the test itself. If there is more than one covariate, these should not be over-correlated.

It has its critics

As is suggested by the above, suitable data sets are quite narrowly defined. While proponents of most parametric tests cite their robustness, there is evidence to suggest that ANCOVA is "a delicate instrument" (Huck 2012). Serious critiques of ANCOVA describe problems relating to data reliability and the smoothing out of differences between mixed groups (Campbell, 1989; Buser, 1995; Miller and Chapman, 2001). Huck (2012) blames the users! He feels that they often consider complexity to be a virtue in itself. His more measured description is thus:

> "To provide meaningful results, ANCOVA must be used very carefully – with attention paid to important assumptions, with focus directed at the appropriate set of sample means, and with concern over the correct way to draw inferences from ANCOVA's F-tests. Because of its complexity, ANCOVA affords its users more opportunities to make mistakes than does ANOVA."

Note that drawing inference, how you interpret the data, is also problematic. Reputable researchers have created serious flaws in their studies by using misleading results from ANCOVA (Campbell, 1989 and Buser, 1995 cite instances from educational research).

Conclusion

This book is aimed primarily at beginners. I would suggest that intermediate, and even advanced statistical users, think carefully before using ANCOVA.

Sequential regression – more words of warning

This comprises a set of algorithms for the automated selection of variables for addition or elimination in regression. In JASP, this is available with both linear regression and logistic regression, using the drop-down menu near the top referred to as 'Method'.

Generally, when using linear regression or logistic regression, I would strongly advise you to leave Method on the default setting, 'Enter'. This introduces all of the variables at the same time.

The other methods – Backward, Forward, and Stepwise – introduce each variable sequentially. These are automated procedures, sometimes referred to collectively as stepwise regression, which is a little misleading, as Stepwise is also one of the procedures. **Backward elimination** starts with all the variables and weeds out the worst predictors early on. **Forward selection** starts without any variables and adds eligible variables. The **Stepwise** option, perhaps better referred to by the less well-known term **Bidirectional elimination**, is similar to Forward selection, but the algorithm revisits the status of existing models at each step and eliminates variables which have become redundant.

There are a lot of criticisms of this overall approach. Perhaps the keystone is that the algorithms are considered to make optimal choices at each step, but poorer choices taken as a whole than other methods. To my mind, this is part of the paradigm of pressing buttons first and worrying about what it all means later (if at all); put another way, the use of these methods can be logically incoherent.

Cohen and Cohen (1983) only recommend these procedures for predictive purposes, as "neither the statistical significance test for each variable nor the overall tests .. at each step are valid". They consider the procedures as usable based on three assumptions: The research purpose is predictive rather than explanatory; there are large samples, with at least 40 cases per predictor; and they have been cross-validated by replication on another sample. The cross-validation is partly because these techniques tend to overfit the models, giving such an accurate portrayal of the sample that they encapsulate the 'noise' of the data as well as useful information.

Thinking point

Is complexity a virtue? Is 'powerful' always a positive or even a meaningful adjective? Will the writer of these comments ever emerge from their cave?

References

Bickel R (2013) *Classical Social Theory in Use.* Charlotte, NC: Information Age Publishing.

Bulman M (2016) Can dental X-rays determine a refugee's age? *The Independent, October 19th.*
<http://www.independent.co.uk/news/uk/home-news/child-refugees-age-can-dental-checks-determine-how-old-someone-is-calais-jungle-a7369531.html >

Bross IDJ (1971) Critical Levels, Statistical Language and Scientific Inference. In Godambe VP and Sprott (eds) *Foundations of Statistical Inference.* Toronto: Holt, Rinehart and Winston.

Buser K (1995) Dangers in using ANCOVA to evaluate special education program effects. At *Annual meeting of the American Educational Research Association.* 18–22 April, San Francisco.

Calkins KG (2005) *An Introduction to Statistics*
<https://www.andrews.edu/~ calkins/math/edrm611/edrm05.htm>

Campbell K (1989) Dangers in using analysis of covariance procedures. At *Annual Meeting of the Mid–South Educational Research Association* (17th, Louisville, KY, Nov 9–11, 1988), ERIC (Educational Resources Center), www.eric.ed.gov

Clark–Carter D (1997) *Doing quantitative psychological research: from design to report.* Hove: Psychology Press.

Cohen J (1988). *Statistical power analysis for the behavioral sciences.* Routledge.

Cohen J and Cohen P (1983) *Applied Multiple Regression/Correlation Analysis for the Behavioural Sciences.* Hillsdale, NJ: Lawrence Erlbaum.

Dallal GE (2012) *Multiple Comparison Procedures.* <www.jerrydallal.com/lhsp/mc.htm>

Darboux JG, Appell PE and Poincaré JH (1908) Examen critique des divers systèmes ou études graphologiques auxquels a donné lieu le bordereau. In *L'affaire Drefus - La révision du procès de Rennes - enquête de la chambre criminelle de la Cour de Cassation.* Ligue francaise des droits de l'homme et du citoyen, Paris, 499-600.

Day RW and Quinn GPB (1989) Comparisons of Treatments After an Analysis of Variance in Ecology. *Ecological Monographs*, 59, 433-63.

Dennis B (1996) Discussion: Should Ecologists Become Bayesians? *Ecological Applications*, 6, 1095-1103. <http://www.webpages.uidaho.edu/~ brian/reprints/Dennis_Ecological_Applications_1996.pdf>

Everitt B, Landau S, Leese M and Stahl, D (2011) *Cluster Analysis.* Oxford: Wiley-Blackwell.

Evett IW (1991) Implementing Bayesian Methods in Forensic Science. *Proceedings of the 4th Valencia International Meeting on Bayesian Statistics.*

Fisher RA (1926), The Arrangement of Field Experiments, *Journal of the Ministry of Agriculture of Great Britain*, 33, 503-513.

Fisher, RA (1935) *The design of experiments.* Edinburgh: Oliver and Boyd.

Games PA (1971) Multiple comparisons of means. *American Educational Research Journal*, 8, 531-565

Gelman A (2011) Induction and Deduction in Bayesian Data Analysis. *Rationality, Markets and Morals, 2* < http://www.stat.columbia.edu/~gelman/research/published/philosophy_online4.pdf >

Girden E (1992) *ANOVA: Repeated Measures.* Sage.

Goodboy AK and Martin MM (2020) Omega over alpha for reliability estimation of unidimensional communication measures. *Annals of the International Communication Association, 44*, 422-439.

Goss-Sampson M (2020) Bayesian inference in JASP. < http://static.jasp-stats.org/Manuals/Bayesian_Guide_v0_12_2_1.pdf >

Greene J and D'Oliveira M (1982). *Learning to use statistical tests in psychology: A student's guide.* Milton Keynes: Open University Press.

Hayes AF and Coutts JJ (2020) Use Omega Rather than Cronbach's Alpha for Estimating Reliability. But... *Communication Methods and Measures, 14*, 1-24.

Hess B and Olejnik S (1997) Top ten reasons why most omnibus ANOVA F-tests should be abandoned. *Journal of Vocational Education Research, 22*, 219-32.

Hilton A and Armstrong RA (2006) Statnote 6: post-hoc ANOVA tests. *Microbiologist, 7*, 34-36.

Hsu J (1996) *Multiple Comparisons.* Chapman and Hall.

Huck, SW (2008) *Statistical misconceptions.* London: Routledge.

Huck SW (2012) *Reading Statistics and Research.* Boston: Pearson.

Jarosz AF and Wiley J (2014) What Are the Odds? A Practical Guide to Computing and Reporting Bayes Factors. *Journal of Problem Solving, 7*.

Jeffreys H (1961) *Theory of Probability* (3rd edition). Oxford: Clarendon Press.

Kim HY (2013) Statistical notes for clinical researchers: assessing normal distribution (2) using skewness and kurtosis. *Restorative Dentistry and Endedontics*, *38*, 52-54.
<http://www.ncbi.nlm.nih.gov/pmc/articles/PMC3591587/>

Kinnear P and Gray C (2004) *SPSS 12 made simple.* Hove: Psychology Press.

Lee MD and Wagenmakers E-J (2013). *Bayesian modeling for cognitive science: A practical course.* Cambridge University Press.

Lucy D (2005) *Introduction to Statistics for Forensic Scientists.* Chichester: John Wiley.

McGrayne SB (2012) *The theory that would not die: how Bayes' Rule cracked the Enigma Code, hunted down Russian submarines, and emerged triumphant from two centuries of controversy.*
New Haven CT: Yale University Press.

Mayo D (2012) *Error Statistics Philosophy.* <errorstatistics.com>

Mead R, Gilmour SJ and Mead A (2012) *Statistical Principles for the Design of Experiments.* Cambridge: Cambridge University Press.

Miller G and Chapman J (2001) Misunderstanding analysis of covariance. *Journal of Abnormal Psychology*, *110*, 40–8.

Morey RD, Hoekstra R, Rouder JN, Lee MD and Wagenmakers E-J (2016) The Fallacy of Placing Confidence in Confidence Intervals. *Psychonomic Bulletin & Review*, *23*, 103-123.

Navarro DJ, Foxcroft DR, Faulkenberry TJ (2019) *Learning Statistics with JASP: A Tutorial for Psychology Students and Other Beginners.*

Nelder J (1971) Discussion on papers by Wynn, Bloomfield, O'Neill and Wetherill. *Journal of the Royal Statistical Society, series B*, *33*, 244-246.

Nelder J (1999) From statistics to statistical science. *Statistician*, *48*, 257–267.

O'Brien RG (1981) A simple test for variance effects in experimental designs. *Psychological Bulletin*, *89* 570–574.

Okada K (2013) Is omega squared less biased? A comparison of three major effect size indices in one-way ANOVA. *Behaviormetrika, 40*, 129–147.

Parker R (1979) *Introductory statistics for biology.* London: Edward Arnold.

Pearce S (1993) Data analysis in agricultural experimentation. III. Multiple comparisons. *Experimental Agriculture, 29*, 1-8.

Pedhazur EJ and Schmelkin LP (1991) *Measurement, Design and Analysis.* Hillsdale NJ: Lawrence Erlbaum.

Perrigot R, Cliquet G and Mesbah M (2004) Possible applications of survival analysis in franchising research. *International Review of Retail, Distribution and Consumer Research, 14*, 129-143.

Plackett, R. (1971) *Introduction to the theory of statistics.* Edinburgh: Oliver and Boyd.

Popper K (1968) *The Logic of Scientific Discovery.* New York: Harper and Row.

Preece D (1982) T is for trouble (and textbooks): a critique of some examples of the paired-samples t–test. *The Statistician, 31*, 169–195.

Raftery AE (1995). Bayesian model selection in social research. In Marsden PV (ed), *Sociological methodology*, 111–196. Cambridge, MA: Blackwell.

Razali NM and Wah YB (2011). Power Comparisons of Shapiro-Wilk, Kolmogorov-Smirnov, Lilliefors and Anderson-Darling Tests. *Journal of Statistical Modeling and Anlytics, 2*, 21-33.

Rice W (1989) Analysing tables of statistical tests. *Evolution, 43*, 223–225.

Sato T (1996) Type 1 and type 2 errors in multiple comparisons. *The Journal of Psychology, 130*, 293-302.

Savage M (2015) *Social Class in the 21st Century.* London: Penguin.

Schonbrodt F (2015) *What does a Bayes Factor feel like?* <http://www.nicebread.de/what-does-a-bayes-factor-feel-like/>

Steinhardt J (2014) *A Fervent Defense of Frequentist Statistics.* <http://lesswrong.com/lw/jne/a_fervent_defense_of_frequentist_statistics/>

Stevens JP (2009) *Applied Multivariate Statistics for the Social Sciences* (5th ed). NY: Routledge.

Swan A (2021) *JASP 0.15 Tutorial: ODDS Ratio in Contingency Tables.* YouTube, November 4, 2021.

Tabachnick BG and Fidell LS (2007) *Using Multivariate Statistics.* Boston: Pearson.

Tsoumakas G, Lefteris A and Vlahavas I (2005) Selective fusion of heterogeneous classifiers. *Selective Data Analysis, 9*, 511–525.

Wagenmakers E (2007) A practical solution to the pervasive problems of p values. *Psychonomic Bulletin & Review, 14*(5), 779-804 <http://www.ejwagenmakers.com/2007/pValueProblems.pdf>

Walsh A and Ellis L (2007) *Criminology.* London: Sage.

Wei Z, Yang A, Rocha L, Miranda MF and Nathoo FS (2022) A Review of Bayesian Hypothesis Testing and its Practical Implementations. *Entropy, 24.*

West SG, Finch JF, Curran PJ (1995) Structural equation models with nonnormal variables: problems and remedies. In Hoyle RH (ed) *Structural equation modeling: Concepts, issues and applications*, 56–75. Sage.

Wyseure G (2003) *Multiple comparisons'* http://www.agr.kuleuven.ac.be/vakken/statisticsbyR/ANOVAbyRr/multiplecomp.htm

Yerkes RM and Dodson JD (1908) The Relation of Strength of Stimulus to Rapidity of Habit Formation. *Journal of Comparative Neurology and Psychology, 18*, 459–482. <https://doi.org/10.1002/cne.920180503>

Index

Page numbers in *italic script* refer to Bayesian tests.

alternative hypothesis, 23, 33, 40, 43, 56
ANCOVA, 210-12
ANOVA (analysis of variance), 64, 176, 177
    factorial, between subjects, 190-93, *193-4*
    factorial, mixed design, 195-8, *199-200*
    factorial, repeated measures (within subjects), 177, 178-83, *183-4*, 185-7, *188-9*
    univariate (1 way), between subjects, 77-84, *84-5*
    univariate (1 way), repeated measures (within subjects), 60-66, *66-7*
autocorrelation, 126-8
Bayesian reporting tables, *40*, *41-2*, *47*, *55*, *99*, *141*, *183*
binomial test, 138-141, *141-3*
Brown-Forsythe correction, 72, 80
censored information, see survival analysis
central tendency, 15-17, 18, 20, 59, 64, 169-71
    mean, 15, 16, 17, 18, 20, 28, 29, 59, 64, 72-3, 98, 123, 170, 204
    median, 15, 16, 17, 59, 73, 170-71
    mode, 16, 17, 170
'chi squared' (chi square test of association / independence), 149-59, *159-60*
chi square goodness of fit, see multinomial

220

cluster analysis, 206-7
collinearity (multicollinearity), 126-8, 168
confidence intervals, 9, 26, 73, 74
correlations, simple, 89-99, *100-101*, 101-104, *105*, 108
correlations, multiple, 108-111, *111-112*, 113-114, *114*, 115
corrections, see multiple comparison tests
Cohen's d, 29, 73
Cramer's V, 154-5
critical value, 24-5, 40, 41, 44, 57, 110-111, 140-41, 171
curvilinearity, 106-107, 127
data types, 14, 18-19
degrees of freedom (df), 154, 169
dispersion, 17-18, 20, 61, 98, 171
Durbin-Watson test, see autocorrelation
effect size, 25, 29, 58-9, 95-6, 108, 177
equality of variances, see homogeneity of variance
eta squared, 65, 87
factor analysis and PCA, 205-206
F ratio, 64, 80, 121, 123, 129, 180
Friedman test, 67-9
gamma, 155
GLM (general linear model), 176
Greenhouse-Geisser and Huyn-Feldt corrections, 197
homogeneity of regression, 211
homogeneity of variance, 19, 80, 196, 204
homoscedasticity, see homogeneity of variance
inferential statistics, 22
Kaplan-Meier survival function / curve, see survival analysis
Kendall's tau b, 97, 101-104, *105*, 113-4, *114*
Kruskal-Wallis test, 86-7
kurtosis, 20, 61, 97, 113-4
Levene's test, 72, 76, 79, 80
life tables, see survival analysis
likelihood function (Bayesian), *37*
linearity, 96, 106-7, 115, 125, 211

logistic (or logit) regression, 207-208
log-linear regression, 161-3, *163-4*
Mann-Whitney test, 72, 75-6, *77*
Mauchly's W (test of sphericity), 64, 197
meta-analysis, 210
multinomial (chi square goodness of fit) test, 143-7, *147-9*
multiple comparison tests, 201-204
    Bonferroni correction, 65, 83, 198, 203, 204
    Conover, 69, 204
    Dunn, 86, 87, 204
    Dunnett, 204
    Games-Howell, 204
    Holm, 202-203
    Scheffe, 203
    Sidak, 204
    Tukey, 203-204
non-parametric tests, 17, 18, 20, 29, 43, 44-5, 58, 59, 67,
    72, 76, 80, 86, 88, 106, 170, 204, 206
normal (Gaussian) distribution, 19-21, 29, 50, 54, 61, 67, 72,
    76, 80, 86, 97, 102, 113-14, 115, 170
null hypothesis, 23-4, 25, 27, 29, 32, 40, 43-4, 45, 80, 94, 95,
    120, 128, 140, 145, 149, 156, 171
Omega squared, 81
outliers / outlying data, 17, 96-7, 121-2, 172
overfitting (of models), 213
parametric tests, 18-20, *43*, 50, 58, 72, 96-7, 113-14, 169, 170, 211
parsimonious model, 128, 184, 189
partial eta squared, 65, 81, 181
Pearson test, 96-9, *99-100*, 108, 108-111, *111-2*
phi, 154
population, 20, 22, 28-29, 31-2, *46-7*, 147, 170
posterior distribution, *36-39*, *56*, *60*, *85*
post hoc tests, see multiple comparison tests
Principal Components Analysis (PCA), see factor analysis
prior distribution, *36-39*

p value, see research design
qualitative analysis, see research design
range, 17, 98
regression, sequential, 212-3
regression, simple, 115-123
regression, multiple, 123-9, *130-133*
reliability, 209-210
research design, 11-14, 48-51, 89-90, 94-5, 114-116, 134, 135-8, 176-7
residuals, 80, 125-6, 127, 128
RMSE (in multiple regression), 119, 128, 129
samples, 20, 22, 25, 28, 31-2, *33-4*, *37*, *43*, *44*, *47*, 73, 81, 97, 101, 128, 153, 213
sequential regression, 212-3
Shapiro-Wilk test, 21, 29, 54, 61, 97
significance, 22-30, 40, 47
skewness, 20, 80, 97, 170-71
Spearman, 97, 101-104, 113-4
sphericity, 64, 180, 196-7
standard deviation, 17, 20, 29, 72-3, 98, 171
survival analysis, 208-09
t test – independent samples (different subjects), 70-74, *74*
t test – one sample, 28-30, *46-7*
t test – paired samples (same subjects), 52-4, *55-6*
type 1 and type 2 errors, 25, 27, 62, 81, 108, 153, 180, 197, 201, 202-203
uninformative priors, *39*
variance, 18, 64
Welch's test, 72, 76, 80
Wilcoxon test, 56-9, *59-60*

www.ingramcontent.com/pod-product-compliance
Lightning Source LLC
Chambersburg PA
CBHW051540020426
42333CB00016B/2023